DEPARTMENT OF GEOLOGY
WHEATON COLLEGE
WHEATON, ILLINOIS 60187

Management of Latin American river basins

D1547084

Water Resources Management and Policy Series Editors

Dr. Juha I. Uitto
Environment & Sustainable Development
United Nations University
Tokyo, Japan

Prof. Asit K. Biswas, President
Third World Centre for Water Management
Mexico City, Mexico

International Advisory Board

Dr. Mahmoud A. Abu-Zeid
Minister of Public Works and Water Resources, Cairo, Egypt

Prof. Benedito P.F. Braga
President, International Water Resources Association
Escola Politecnica, University of São Paulo; São Paulo, Brazil

Dr. Ralph J. Daley
Director, UNU/INWEH; Hamilton, Canada

Dr. David Seckler
Director-General, International Water Management Institute; Colombo, Sri Lanka

Dr. Ismail Serageldin
Vice-President, The World Bank; Washington, D.C., USA

Dr. Aly M. Shady
Canadian International Development Agency; Hull, Canada

Prof. Yutaka Takahasi
Professor Emeritus, Tokyo University

Dr. José Galizia Tundisi
President, CNP_q; Brasilia, Brazil

The UNU Programme on Integrated Basin Management focuses on water management, approaching the complex problematique from three particular angles: governance, capacity-building, and management tools. The programme is carried out through field-based research encompassing both natural and social sciences. It utilizes extensive networks of scholars and institutions in both developing and industrialized countries. This work is intended to contribute to policy-making by the United Nations and the international community, as well as to capacity-building in developing countries.

The Water Resources Management and Policy publications series disseminates the results of research carried out under the Programme on Integrated Basin Management and related activities. The series focuses on policy-relevant topics of wide interest to scholars, practitioners, and policy-makers.

Earlier books in this series are *Hydropolitics Along the Jordan River: Scarce Water and Its Impact on the Arab-Israeli Conflict*, by Aaron T. Wolf; *Managing Water for Peace in the Middle East: Alternative Strategies*, by Masahiro Murakami); *Freshwater Resources in Arid Lands*, edited by Juha I. Uitto and Jutta Schneider; and *Central Eurasian Water Crisis: Caspian, Aral, and Dead Seas*, edited by Iwao Kobori and Michael H. Glantz.

Management of Latin American river basins: Amazon, Plata, and São Francisco

Edited by Asit K. Biswas, Newton V. Cordeiro, Benedito P.F. Braga, and Cecilia Tortajada

United Nations University Press

TOKYO · NEW YORK · PARIS

© The United Nations University, 1999

The views expressed in this publication are those of the authors and do not necessarily reflect the views of the United Nations University.

United Nations University Press
The United Nations University, 53-70, Jingumae 5-chome,
Shibuya-ku, Tokyo 150-8925, Japan
Tel: (03) 3599-2811   Fax: (03) 3406-7345
E-mail: sales@hq.unu.edu; http://www.unu.edu

United Nations University Press is the publishing division of the United Nations University.

Cover design by Joyce C. Weston

Cover photograph by Newton V. Cordeiro

Printed in the United States of America

ISBN 92-808-1012-X

Library of Congress Cataloging-in-Publication Data

Management of Latin American river basins : Amazon, Plata, and São Francisco / edited by Asit K. Biswas ... [et al.].
    p. cm.—(Water resources management and policy series)
  UNUP-1012"-T.p. verso.
  Includes bibliographical references and index.
  ISBN 928081012X (pbk.)
  1. Water-supply-Latin America-Management-Congresses.
2. Water resources development-Amazon River Watershed-
Congresses. 3. Water resources development-Rio de la Plata
Watershed (Argentina and Uruguay)—Congresses. 4. Water
resources development-São Francisco River Watershed
(Brazil)—Congresses. 5. Water resources development-
International cooperation-Congresses. 6. Sustainable
development-Latin America-Congresses. I. Biswas, Asit K.
II. Latin American Water Forum (1997:São Paulo, Brazil)
III. Series.
  TD227.5 .M36 1998
  333.91'15'098—dc21   98-58084
                                                       CIP

This book is dedicated to Reizo Utagawa, Managing Director, Nippon Foundation, as a token of esteem for a great statesman and journalist and as a mark of true regard for a friend

Contents

Foreword

With the end of the Cold War and the resulting waning of East-West tensions, a new type of international and regional security, which is non-military in character, is becoming an increasing international concern. The conventional military dimensions of peace and security aspects still persist, though at a much lower level of importance than before. This lowering, however, has been more than compensated by increasing unconventional security threats from new factors like population growth and the associated impacts such as depletion and degradation of natural resources and environmental deterioration.

On a historical basis, these new factors did not generally have direct and immediate linkages to "traditional" peace and security issues, as exemplified by the preponderance of the military considerations during the Cold War. However, if these emerging issues continue to be neglected, or given inadequate attention, as mostly is the case at present, they are likely to intensify, and thus contribute to serious threats to national and regional peace in the coming decades. Such neglects would also multiply significantly the number of potential threats that are likely to be witnessed on a global basis.

Very few people have realized the importance of these new and emerging factors as a serious threat to future national, regional, and global peace, and even fewer have a clearer understanding of where, how, and when such threats could affect peace and security issues. To the extent these non-conventional concerns have received some attention in recent years, they have generally tended to concentrate only on one issue: global warming and the associated climate change. In my view, the most immediate and the most serious threat to national and regional peace and security is likely to come *not* from climate change, though it is an important issue, but from the lack of good quality freshwater. Unfortunately, this is a fact that has been mostly neglected in recent global discussions.

As Professor Asit K. Biswas, the convenor of the Latin American Water Forum, has pointed out elsewhere, climate change has not killed a single person, and is unlikely even to kill a single person during the coming two decades. Yet, climate change has received almost all the global attention in this area thus far. In contrast, absence of clean water, scarcity of water, and floods are killing more than six million people each year, and this fact has been virtually ignored in the international political agenda and media coverages. An objective review and a comprehensive global analysis of population growth, resource depletion, and environmental degradation would indicate that the scarcity of good quality water would most likely be the next major crisis facing the world in the early part of the twenty-first century. Unless necessary countermeasures are taken in the near future, water crisis could trigger national and regional instabilities in many parts of the world in the coming years.

The central role water plays in sustaining all life on earth and preserving the various ecosystems is beyond dispute. As the populations in developing countries continue to increase, more and more water will be necessary for human consumption, industrial expansion, electricity generation, food production, and ecosystem preservation. Since the developing countries are almost exclusively located in the tropical and semi-tropical regions, water has already become a scarce commodity in many such countries. As the demands for fresh water continue to increase in all countries in the future, national and regional tensions over the use of water sources that could be economically and environmentally developed will become more and more intense. Specifically, the problems associated with the use of international water bodies, that is those rivers, lakes, and aquifers that are shared by two or more countries, would become increasingly complex, acute and fraught with danger. The ever-increasing competition for limited supplies of fresh water between neighbouring countries could lead to severe political tensions, and even to war under certain special conditions.

Tensions over the use of international water bodies have been serious in the past. For example, an important contributory cause for the Arab-Israeli War of 1967 was the struggle over the control of the Jordan river and the other water bodies of the region. Long-standing disputes over the water use of the Rio Lauca has led Bolivia to sever diplomatic relations with Chile. Similarly relations between Bangladesh, India, and Nepal over the use of the waters of the River Ganges, between Syria and Turkey over the Euphrates river, and between Egypt and Ethiopia over the River Nile, have been strained in the past. Former Foreign Minister of Egypt, Boutros Boutros-Ghali, who later became the Secretary General of the United Nations has said that the "next war in our region will be over the waters of the Nile, not politics."

Fortunately, all water problems take some time before they reach critical proportions. In the past, in a few cases appropriate timely actions were taken which prevented avoidance of deadly conflicts later. A good example is the agreement between India and Pakistan over the Indus river. This 1960 agreement was made possible by the dynamic leadership shown by Eugene Black, the then President of the World Bank. The fact that the World Bank successfully played the role of an honest broker, and that it was willing to provide substantial economic assistance to both the countries if and when a mutually acceptable treaty on the Indus System could be signed, greatly facilitated the negotiating process. The Bank later provided $720m in economic assistance to the two countries concerned. Regrettably however, the leaders of international organizations have become increasingly risk-averse, and the type of leadership that was shown by Eugene Black has been conspicuously absent during the past 35 years.

There are many international rivers where problems have still not reached critical proportions. I strongly believe that by raising the alarm objectively and forcefully today, it is possible to increase the awareness of governments, people of the river basins, and international organizations, in terms of the importance and the criticality of the issues involved. Such a process could contribute to the emergence of the necessary political wills in the nations concerned so that the problems could be resolved before they become increasingly more critical, complex, and entrenched.

This is the main reason for the Nippon Foundation's decision to support the Latin American Water Forum, which critically analysed the inter-state and intra-state problems of the Amazon, Plata, and São Francisco rivers.

Professor Biswas, the convenor of the Forum, using his extensive personal contacts and influence in the region, brought together very high ranking decision makers and selected technocrats from all the countries associated with the three rivers, as well as the major international organizations. Such an important event had simply not occurred earlier. The present book includes the edited discussion papers especially commissioned for this very successful Forum. It is my fervent hope that the results of the Forum will go a long way towards the timely development of efficient frameworks which will contribute to the sustainable use of the waters of these three important Latin American rivers, and thus help to prevent any future potential conflicts.

Reizo Utagawa

Reizo Utagawa
Managing Director, Nippon Foundation, Tokyo, Japan

Preface

It was Leonardo da Vinci who said water was the "driver of nature." Some may have considered this to be an overstatement in the earlier era, but as the twentieth century draws to a close, the relevance and importance of Leonardo's view is becoming abundantly evident, even though it was made several centuries ago.

Increasing population, higher levels of human activities and accelerating societal concern for preserving the environment have made sustainable water management an increasingly difficult and complex task throughout the world in recent years. The problem is being further compounded by the fact that the per capita demand for water in nearly all developing countries is steadily increasing as more and more people reach higher standards of living, and as lifestyles in developed countries continue to change. The situation is becoming increasingly critical in nearly all developing countries, where the rates of population growth are the highest, and unlike in developed countries, water scarcities have very significant impacts on national economies.

Furthermore, most of the economically exploitable sources of water in developing countries have already been developed or are in the process of development. Current analyses indicate that the cost of developing each cubic metre of water for the next generation of water projects is likely to be two to three times higher than the present generation in real terms. This is a fact that thus far has been basically ignored by water planners all over the world.

It is now becoming increasingly evident that the water management profession will face a problem during the early part of the twenty-first century, the magnitude and complexity of which no earlier generation has had to face. In the twilight of the twentieth century, the water profession faces two stark and radically different choices: to carry on as before with a "business as usual" approach which contributes primarily to only incremental changes, and thus endow our future generations with increasing water scarcities and extensive water pollution problems, or earnestly pursue an accelerated effort which could contribute to fundamental changes in our mindsets as to how the world's water resources are to be planned, managed, and used.

Latin America is no exception to this overall global trend. Thus, not surprisingly, at the Bolivia Summit of the Americas, which was held in December 1996, the leaders of all the countries of this region categorically declared that:

Despite extensive efforts by countries in the Americas to improve water use and management, demand continues to rise while contamination has seriously degraded the quality of freshwater, spreading disease and causing economic losses.

Poor management structure and pricing, as well as lack of stakeholder commitment to water management and conservation, are important factors contributing to growing scarcity. Particularly troublesome are the projected demands of drinking water by urban populations, and potential conflicts among sectors, regions, and countries that share water resources.

The leaders then went on to identify transboundary water conflicts as one of the main priority areas of concern.

Faced with the above facts and trends, the Committee on International Collaboration of the International Water Resources Association (IWRA) decided to convene a Latin American Water Forum in São Paulo, Brazil, 15–17 January 1997. The Forum considered two major international river basins: Amazon (shared by eight countries: Brazil, Peru, Bolivia, Colombia, Ecuador, Guyana, Suriname, and Venezuela; the countries are listed in descending order in terms of share of the river basin areas they contain), and Plata (five countries,

Brazil, Argentina, Paraguay, Bolivia, and Uruguay). In terms of the size of the river basins, Amazon (7,800,000 km^2) and Plata (3,100,000 km^2) are the two largest river basins in Latin America.

The Forum also considered a third river, São Francisco, which is one of the major rivers of South America, even though it is totally within the Brazilian territory. The São Francisco river basin covers a drainage area of 640,000 km^2, and it provides a lifeline to the arid north-east region of Brazil, which is nationally known as the drought polygon. Since the river passes through numerous Brazilian states, inter-state conflicts are likely to become increasingly more serious in the future. While much has been written on conflicts between countries sharing international rivers, inter-state conflicts on exclusively national rivers can sometimes be more intense and complex. For example, the time required to reach an agreement between the four provinces of Pakistan to share the water of the Indus System was more than three and a half times longer than was required to negotiate a treaty between India and Pakistan. Also, more people have died because of conflicts on inter-state rivers as compared to those on international rivers.

Participation in the Forum was restricted to 33 well-known experts of Latin America. They were carefully selected and invited because of their acknowledged expertise in the field. All participants were invited in their personal capacities for a free and frank exchange of ideas, opinions, and facts.

The Latin American Water Forum is the third regional event that the IWRA Committee on International Collaboration has convened in recent years. The first two were the Middle East Water Forum (Cairo, Egypt, 7–9 February 1993) and the Asian Water Forum (Bangkok, Thailand, 30 January-1 February 1995). Both the Cairo and Bangkok Forums were co-sponsored by the United Nations University and the United Nations Environment Programme, and resulted in two definitive books: *International Waters of the Middle East: From Euphrates-Tigris to the Nile* (Oxford University Press, 1994) and *Asian International Waters: From Ganges-Brahmaputra to Mekong* (Oxford University Press, 1996). The Cairo Forum also received considerable support from the Sasakawa Peace Foundation.

The Cairo Forum directly contributed to the establishment of the Middle East Water Commission, which I also had the honour and privilege to chair. The Commission was supported by the Sasakawa Peace Foundation. Its authoritative report *Core and Periphery: A*

Comprehensive Approach to Middle Eastern Water was published by Oxford University Press in 1997. An Arabic edition of this book is available from Dar An-Nahar Publishing, Beirut, Lebanon.

A Japanese version of the *Asian International Waters* book is available from Keiso Sholbo, Tokyo, Japan. The Bangkok Forum was followed by a very high level Ganges Forum in Calcutta, India, 18–20 March 1998. The book based on the Ganges Forum will be published shortly by the United Nations University Press.

An event like the Latin American Water Forum cannot be organized without the support of many individuals. First and foremost, I am most grateful to Mr. Reizo Utagawa, Managing Director of the Nippon Foundation, Tokyo, and Dr. Takashi Shirasu of the Sasakawa Peace Foundation, Tokyo. Both Utagawasan and Shirasusan not only strongly encouraged me to proceed with the Forum, but also gave me excellent advice during my regular visits to Tokyo. The Nippon Foundation also provided financial support to the Forum, and Utagawasan kindly consented to write a Foreword for this book. As a small token of our appreciation, this book is dedicated to Utagawasan. I would also like to express our thanks to Mr. Yasuhisa Yamada and Ms. Eriko Anraku of the Nippon Foundation for their continued support throughout this project.

I am also most grateful to Newton V. Cordeiro, formerly of the Organization of the American States. Right from the very beginning, Dr. Cordeiro took a keen personal interest in the Forum. His extensive personal contacts in Latin America were instrumental in ensuring a very high level participation of decision makers from the co-basin countries of the three rivers. He assured OAS financial support for the Forum, and also wrote one of the major papers. To a very significant extent, the success of the Forum was due to the strong support and excellent advice I received on a regular basis from Dr. Cordeiro.

I would also like to express our most sincere appreciation to Professor Jose Galizia Tundisi, President of the Conselho Nacional de Desenvolvimento Cientifico e Tecnológico (CNPq) of Brazil. Professor Tundisi, an eminent water scientist and a friend of long standing, not only gave the key-note lecture to the Forum but also provided financial assistance through CNPq.

It should be further noted that without the strong support of Professor Benedito P.F. Braga, President of the International Water Resources Association, the Forum simply could not have been convened in São Paulo. He wrote one of the main papers of the Forum, and was also responsible for the entire organizational arrangements of the Forum.

Last but not least, I wish to acknowledge the assistance of Cecilia Tortajada, who undertook the lion's share of the editing work of this book. On behalf of the IWRA Committee on International Collaboration, I gratefully acknowledge the support and assistance of all the above individuals which were primarily responsible for making the Forum a great success.

Like the earlier two Forums, our Committee is now planning follow-up activities for the São Paulo Forum. Those who are interested can check the latest plans at our website www.iwra.siu.edu

Asit K. Biswas

Asit K. Biswas, Chairman,
IWRA Committee on International Collaboration,
Mexico City, Mexico

Part I
The Amazon river basin

1

Sustainable water-resources development of the Amazon Basin

B. Braga, E. Salati, and H. Mattos de Lemos

Introduction

The Amazon region includes the greatest area of tropical rain forests and the largest river basin in the world. Contrary to popular belief, the region comprises a great number of different ecosystems, with varied geological, geomorphological, soil and climatic characteristics, resulting in a highly diversified flora and fauna. In spite of its immense natural resources – huge amounts of wood, water, and rich mineral deposits – scientists today are convinced that its greatest value lies in the vast biodiversity and the potential locked up within it. In the Brazilian Amazon, the approximate value of the known mineral resources (iron, bauxite, copper, gold, manganese, nickel, silver, and tin) have been estimated at $1,600bn (Comision Amazonica de Desarrollo y Medio Ambiente, 1994). The value of the biodiversity of the region is not known yet, although the market value of the existing commercial wood was estimated at $1,700bn (Mattos de Lemos, 1990).

A 1982 US National Academy of Sciences report estimated that a typical 4-square-mile patch of rain forest may contain 750 species of trees, 125 kinds of mammals, 400 types of birds, 100 of reptiles, and 60 of amphibians. Each type of tree may support more than 400 insect species. Just for comparison, the temperate forests of France contain only about 50 species of trees. As one of the last unexplored regions of the earth, the Amazon exerts much fascination over the imagination of mankind, especially among those who live in less luxuriant ecological settings (Repetto, 1988). Since the fall of the myth which professed the Amazon to be the "lungs of the world," scientists have attempted to understand its precise influence on the maintenance of the global climate system. Roughly half of the rain that falls in the Amazon is generated inside the region, resulting in a hot and very humid climate throughout most of the basin.

The largest part of the Amazon basin is a plain below 200 m above sea level, more than 3,400 km wide from east to west and 2,000 km long from north to south. This large plain is surrounded in the north by the Guyana plateau, at 600–700 m, in the south by the Brazilian plateau with an average height of 700 m, and in the west by the Andes mountain range rising to heights above 4,000 m. The main river system, namely Amazonas Solimões-Ucayali, extends for 6,671 km, which is 91 km longer than the Nile river – once considered the longest in the world. The Amazon river, with more than 1,000 tributaries, discharges into the Atlantic ocean between 200,000 and 220,000 cubic metres per second – about 60 times the rate of the Nile. The discharge of the Amazon river is equivalent to 15.47 per cent of all the fresh water entering the oceans each day (Mattos de Lemos, 1990). Although the gradient of the river is very pronounced in the Andean region, from the foothills to the estuary, the gradient drops only 107 m between Iquitos in Peru and the estuary in Brazil over 2,375 km. The water level varies considerably during the year from 6–10 m near the mouth to between 10 and 15 m in the middle stretch.

The high dependency of the population and the region's economy on the waters of the river, in terms of fresh water for domestic use, fishing activities, transportation, and energy generation, makes protection of the rivers and lakes a top priority – particularly against pollution, overfishing, erosion, mangrove destruction, and drainage of marshy areas. The water quality in the basin is still very good (the amount of dilution is enormous). New techniques for the recovery of the used mercury in gold mines in the Brazilian Amazon are underway. In this huge waterway live approximately 2,000 species of fish, more than all the aquatic fauna found in the Atlantic ocean. The quantity of animal protein from these fish that can be sustainably exploited is potentially enormous. This could provide a sensible

alternative to efforts to produce animal protein from cattle raising –
an activity extremely destructive to the Amazon ecosystem.

Even though the Amazon region has more than 40,000 km of roads
(the majority are still unpaved), the rivers continue to be the main
corridors for transportation with several thousand km of waterways
navigable more than 90 per cent of the time. The alternatives for
energy production using resources of the region include hydropower,
fossil fuels (oil and gas), biomass, and solar energy. With respect to
hydropower, the best sites are in the surrounding mountains, especially on the Andean side. Important hydropower reserves can also
be found on the Amazonian plain, where Brazil has about 45 per cent
of its hydropower potential. However, the reservoirs for this purpose
cover extensive areas, and therefore their proposed construction has
met strong opposition from environmentalists.

This paper describes the Amazonian basin from a physical and
ecological point of view and indicates the current tendencies for
sustainable use of its water resources. It has a first chapter on the
natural system where the hydrology, meteorology, and ecology of
the region is discussed. The second chapter deals with the water use
for hydropower and navigation in the basin and the third chapter
describes the legal and institutional arrangements in this basin at the
Brazilian national level and in the realm of the Treaty for Amazonian
Cooperation.

The natural system

General description

The so-called "Amazon region" extends its limits beyond the Amazon river basin. This "Amazonian dominion" (figure 1) comprises
nine South American countries including, Bolivia, Brazil, Colombia,
Ecuador, Guyana, French Guyana, Peru, Venezuela, and Suriname
over an area of approximately 7.5 million km². Besides the Amazon
river basin this dominion includes part of the Tocantins and Orinoco
river basins and some small basins draining directly to the Atlantic.
The Brazilian Amazon represents approximately 50 per cent of the
country. In 1966 a presidential decree instituted the Legal Amazon
which encompasses today eight states (Acre, Amazonas, Pará, Ror-
aima, Amapá, Rondônia, Mato Grosso and Tocantins) and part of
the state of Maranhão. The population of this area is 18 million dis-

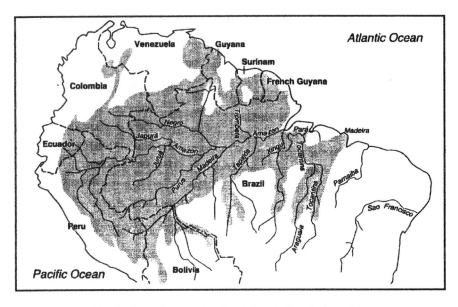

Fig. 1   **The Amazonian dominion in South America**

tributed among 629 municipalities with a progressively high urban concentration (approximately 60 per cent).

The total drainage area of the Amazon river basin is 6,112,000 km^2. The average annual precipitation is of the order of 2,460 mm/year and the average annual flow 209,000 m^3/s which implies a specific flow of 34.2 l/s/km^2. The average annual evapotranspiration totals 1,382 mm/year. As it can be appreciated in figure 2 this is by far the largest basin in the world in terms of discharge. Some of the tributaries of the Amazon are among the longest rivers in the world. Inside the Brazilian territory the Amazon basin occupies 3,900,000 km^2, with an average annual precipitation of 2,220 mm/year, a specific flow of 30.8 l/s/km^2 and an evapotranspiration of 1,250 mm/year.

The Tocantins river basin, which is partially inside the Amazonian dominion, has an area of 557,000 km^2 and an average precipitation of 1,660 mm/year. The flow at the mouth is 11,800 m^3/s with a specific flow of 15,6 l/s/km^2. The evapotranspiration is in the order of 1,168 mm/year. The rivers from the Amapá state, draining to the Atlantic ocean, between the Oiapoque and the Araguari river basin, comprise an area of 76,000 km^2. The precipitation in this region is of 2,950 mm/year, being the total flow in this stretch of 3,660 m^3/s. The specific

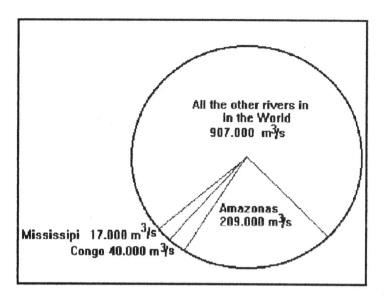

Fig. 2   **Comparative chart for average flow of Amazon river with other rivers in the world (Salati et al., 1983)**

flow is of 48.2 l/s/km². The evapotranspiration is in the order of 1,431 mm/year.

While being a region with a predominantly hot and humid climate, and having a vegetal cover of primarily forest, the Amazon is far from being a natural homogeneous region. Recent studies, taking into account both the biotic and the abiotic characteristics of the ecosystems, have identified 104 systems inside the Amazon, according to the classification of the Natural Landscape Systems (IBGE, 1995). Soils are particularly of low fertility, but are generally covered with a biota extremely rich in species, showing one of the largest natural biodiversities on the planet. The low soil fertility, particularly over the so-called A stable ground surface, together with the climatic conditions, have turned out to be one of the main difficulties for the implementation of a colonization based on agricultural activities.

Although there is little scientific and technical knowledge about the region, large advances have been attained, particularly in the past 30 years with the creation of some specialized institutions such as the National Institute for Research of the Amazon, in Manaus and the State of Pará Emilio Goelde Museum, in Belém. In the past 30 years the number of published works about the Amazon has been growing exponentially, addressing both the biotic and the abiotic aspects. Many of those have been spread out through an incredible number of publications, calling for some efforts to compile and systematize the available information in the form of books and technical journals. Among the scientific journals can be mentioned the *Acta Amazônica*, the *Amazoniana* and the *Museu Goelde Bulletin*. Some of the books presenting extracts from publications are the following: *Amazônia: Development, Interaction and Ecology*, by E. Salati et al. (1983); *The Amazon*, edited by Herald Sioli (1984); and *Amazonian Deforestation and Climate*, edited by J.H. Gash et al. (1996).

Human occupation of the Amazon Region after its discovery by the Europeans can be divided into three phases. The first phase is from 1500 to 1840 and corresponds with the recognition and occupation of the territory. During this period, environmental impact was small, but the interaction of the European settlers with the Indians produced a drastic reduction of native tribes, particularly along the rivers. It has been estimated that more than 3 million Indians lived in an area corresponding to the present Brazilian Amazon. This population was reduced to approximately 160,000 individuals by the second half of the twentieth century. The second phase of colonization covers the period 1840–1912, and is characterized by a drastic increase of rubber exploitation and other plant cultivation such as chestnut trees, as well as hunting and fishing for a few species of animals and fishes. The economic centres of Manaus and Belém experienced considerable growth during this period, and the Brazilian Amazon region received 600,000–800,000 immigrants, particularly from the Brazilian north-east.

In the past five decades a period of modern and intensive colonization can be identified. Roads were built, crossing the region from east to west and from north to south (figure 3). The four-century riparian occupation, particularly along large rivers, expanded inland through the so-called "terra-firma" ecosystem, causing a large environmental and cultural impact on the indigenous communities living in these areas. This wave of modern colonization brought about an incredible increase of deforestation activities. The cleared land, which was only 0.5 per cent of the Amazon region by 1970, expanded to 469,978 km^2 or 9.4 per cent of the Legal Amazon by August 1994 (Barbosa, 1996). The transformation forces prevailing during those decades are associated with the exploitation and use of the natural resources, renewable and non renewable. Main activities were associated with road and hydroelectric plant construction, oil extraction,

Fig. 3   **Highways crossing Brazil and other countries in South America (Salati et al., 1990)**

9

exploitation of forest and fishing resources, cattle raising, tourism and commerce, particularly following the installation of the duty free Manaus zone.

Ecological characteristics of the region

Geomorphology

Over most of its extensive area, the Amazon river basin is formed by geological clusters with altitudes below 250 m above sea level. These lowlands are limited in the west by a semi-circle formed by the Andes mountains with altitudes of over 4,000 m. To the north they are limited by the Amazon residual plateau (Guyana plateau) with an average altitude of only 1,200 m, but showing high summits such as the Neblina peak of 3,014 m. To the south, the Amazon plateau is limited by the Brazilian central plateau with average altitudes varying from 100–400 m. In this way, the Amazon basin has the general format of a horse-shoe with the open side turned to the Atlantic ocean. This physical structure and its geographic position, crossed by the equator, has established important regional characteristics. The water vapour carried from the Atlantic ocean by the trade winds (hot and humid) and the abundant solar energy determine the climate prevailing in the region, with high rainfall rates.

Recent studies have shown that the Amazon river starts at the Nevado de Misme, a mountain in the south of Peru. The waters spring up from the north side of the Chila mountain range in a slope called "quebrada" Carhuasanta, 5,300 m above sea level. The main contributor to the Amazon river, the Apurimac – Ucayalí runs towards the north and, after joining the Marañon river, to the east. Subsequently it takes in water from the Napo river just before crossing the frontier between Peru and Brazil, and is then known as the Solimões river. In Manaus the Solimões takes in the Negro and, from there to its mouth is called the Amazon river. Nowadays, the Amazon river is considered the biggest river in the world, not only in terms of its water volume but also in length, slightly surpassing the Nile at 6,671 km. The real length of the Amazon river is still unknown due to difficulties associated with the exact location of its mouth.

It is important to stress how small is the slope of the main course of the Amazon river; from Iquitos, Peru, to the Marajó island estuary, the river runs for 2,375 km with a total drop of only 107 m. However, an important feature of this relatively flat area is that it is cut by several tributaries, which have carved deep channels with steep slopes at 45 degrees. Then, when considering the large amount of flat land below the forest, what is actually found is a scarred area, impairing the implementation of agricultural activity and highly susceptible to erosion when the original vegetal cover is removed.

Climate

Solar energy

The quantity of solar energy reaching the upper atmosphere in the Amazon region varies very little, since it belongs to the equatorial region. By taking the city of Manaus as an example, we will verify that the maximum of solar energy is equivalent to 885 calories per cm^2/day in January with a minimum of 730 calories in June. In the same way, the length of the diurnal period has also a short variation, going from a minimum of 11.36 hours to a maximum of 12.38 hours. The available energy at ground or at tree top level is controlled by the nebulosity, which is very large in most of the Amazon region, particularly along its central part. The average insolation and the nebulosity are shown in tables 1 and 2, and the incident energy, measured in Manaus, is in table 3, for years 1977–1979.

The most important factor for heat balance is associated with the high nebulosity index and the high values of air humidity. As a result, the temperature remains quite stable with little variation on either daily or annual average.

Air temperature

One of the important characteristics of the Amazon region is the small temperature variation, particularly in lower altitude areas forming the large Amazon plateau. In the city of Belém, for example, the highest monthly temperature is 26.9°C, occurring in November and the lowest is 24°C, in March. In Manaus, the highest monthly average is 27.9°C and the lowest is 25.8°C, both occurring in September. In the city of Iquitos, Peru, the highest monthly average is 32°C occurring in November and the lowest is 32°C, in July. This isothermical condition is a consequence of the large quantity of atmospheric water vapour and the small variation in available solar energy during the year. Table 4 shows the monthly average temperatures occurring in some of the Brazilian Amazon cities.

Table 1 **Mean daily hours of sunlight measured in Amazon cities (hours and minutes)**

N (OMM)	Stations	Lat.	Long.	Jan.	Feb.	Mar.	Apr.	May	June	July	Aug.	Sept.	Oct.	Nov.	Dec.
82,067	Iauareté	0°36′N	69°12′W	4:24	3:54	3:54	3:30	3:36	3:42	3:48	4:42	5:12	4:48	4:47	4:24
82,331	Manaus	3°08′S	60°01′W	3:48	3:55	3:36	3:48	5:24	6:54	7:54	8:12	7:29	6:36	5:54	4:54
82,106	Uaupés	0°08′S	67°05′W	5:13	5:30	5:13	4:31	4:54	4:54	5:13	6:00	6:36	6:07	6:00	5:30
82,741	Taperinha	2°24′S	54°51′W	3:12	3:25	3:24	4:11	6:05	7:49	8:30	8:18	5:43	5:05	4:12	3:29
82,191	Belém	1°28′S	48°29′W	5:01	4:00	3:17	4:13	6:18	7:55	8:37	7:31	7:48	7:55	7:18	6:49
82,243	Santarém	2°45′S	54°42′W	4:35	3:43	3:25	3:48	4:42	5:54	6:49	7:49	7:24	7:24	6:30	6:05

INEMET (1979).

Table 2 Mean cloud cover (in tenths of sky covered)

N (OMM)	Stations	Lat.	Long.	Jan.	Feb.	Mar.	Apr.	May	June	July	Aug.	Sept.	Oct.	Nov.	Dec.	Average
82,704	Cruzeiro do Sul	7°38'S	72°36'W	8.4	8.5	8.4	8.0	7.4	6.5	5.7	5.6	6.9	7.8	8.0	8.2	7.4
82,113	Barcelos	0°58'S	62°57'W	7.1	7.2	7.1	7.7	7.7	7.2	6.7	6.3	6.5	6.8	6.4	6.8	7.0
82,425	Coari	4°05'S	63°08'W	6.7	6.7	6.8	6.7	6.4	5.6	5.2	5.1	5.4	6.0	6.0	6.3	6.1
82,212	Fonte Boa	2°32'S	66°01'W	7.3	7.1	7.2	7.7	7.2	6.9	6.8	6.3	6.3	6.6	6.7	7.0	6.9
82,727	Humaitá	7°31'S	63°02'W	7.5	7.6	7.4	7.0	5.6	3.9	3.1	3.3	5.0	6.2	6.6	7.0	5.8
82,067	Iauaretê	0°36'N	69°12'W	7.6	7.7	8.0	8.2	8.0	7.9	7.7	7.4	7.3	7.7	7.5	7.5	7.7
82,331	Manaus	3°08'S	60°01'W	8.3	8.4	8.5	8.5	7.9	6.9	6.4	6.1	6.9	7.6	7.8	8.0	7.6
82,103	Taracuá	8°10'S	70°46'W	7.3	7.4	7.2	7.6	7.6	7.6	7.4	7.0	6.6	7.2	7.2	7.1	7.3
82,106	Uaupés	0°08'S	67°05'W	8.0	7.8	7.9	8.2	8.1	8.0	7.7	7.4	7.1	7.7	7.6	7.7	7.8
82,741	Taperinha	2°24'S	54°41'W	8.2	8.3	8.2	7.5	5.9	3.6	2.9	3.6	6.4	7.4	7.7	8.1	6.5
82,191	Belém	1°28'S	48°29'W	7.7	8.3	8.6	8.2	7.4	6.1	5.6	5.2	5.6	5.5	6.0	6.8	6.8
82,243	Santarém	2°45'S	54°42'W	6.5	7.0	7.2	7.1	6.5	5.4	4.5	3.7	3.9	4.5	5.0	5.5	5.6

INEMET (1979).

Table 3 Global radiation at surface (Q$_s$) in Manaus; measurements made by Eppley pyranometer (in cal.cm^{-2}.day^{-1})

Year	Jan.	Feb.	Mar.	Apr.	May	June	July	Aug.	Sep.	Oct.	Nov.	Dec.
1977	–	–	–	–	–	–	–	425	348	325	360	267
								(51)	(40)	(36)	(41)	(30)
1978	295	277	305	323	335	432	404	462	486	499	407	363
	(33)	(31)	(34)	(38)	(42)	(57)	(52)	(56)	(56)	(56)	(46)	(41)
1979	337	395	412	–	–	–	–	–	–	–	–	–
	(38)	(36)	(44)									

Note: The numbers in brackets show the percentage in relation to solar radiation reaching the top of local atmosphere (Q$_o$).
Ribeiro et al. (1982).

Precipitation

Rainfall is quite irregular in the Amazon region, as far as spatial distribution is concerned. Figure 4 shows the monthly variation from 34 meteorologic stations. It is immediately obvious that there is a difference in the distribution of monthly precipitation between stations located on the south side of the equator and ones located on the north. The south stations show a dry period from May to August, while the ones from the north indicate, in the same period, the maximum values of precipitation. Another important observation is that precipitation increases going westward, reducing, in the same way, the length of the dry periods. In the north-east of the Amazon basin there is almost no drought through the year.

From the point of view of total precipitation, the minimum values observed for the Amazon basin are in the order of 1,600 mm, along the transition area towards the central Brazilian plateau, and the maximum occur on the slopes of the Andes mountains, showing precipitation above 6,000 mm/year. Precipitation of over 3,000 mm can also be observed on the shores of Amapá and the northern region of the Marajó island. Figure 4 indicates the rainfall distribution and Figure 5 the lines of equal monthly average precipitation (isohyets).

Water balance – origin and recirculation of water vapour in the Amazonia

The water balance area of the Amazon basin is of 6,112,000 km^2. The average yearly precipitation is 2,400 mm/year and the flow of the Amazon river is 209,000 m^3/s. The evapotranspiration corresponds then, to a value of 1,382 mm/year. These are the data published by

Table 4 **Mean temperatures for several regions of the Amazon river basin**

Station	Lat.	Long.	Jan.	Feb.	Mar.	Apr.	May	June	July	Aug.	Sept.	Oct.	Nov.	Dec.	Average
Taperinha-Pa	2°24'S	54°41'W	25.2	25.3	25.2	25.4	25.5	24.8	24.4	25.6	25.8	25.7	25.3	25.4	25.3
Altamira-Pa	3°12'S	52°12'W	25.3	26.6	25.3	25.8	25.8	26.3	25.5	26.1	26.4	26.6	26.4	26.2	26.0
Arumanduba-Pa	1°29'S	52°29'W	27.0	26.8	26.8	26.8	27.2	27.0	26.8	27.0	27.4	27.6	27.9	27.3	27.1
Belém-Pa	1°27'S	48°29'W	25.6	25.5	25.4	25.7	26.0	26.0	25.9	26.0	26.0	26.2	26.5	26.3	25.9
Cachumbo-Pa	9°25'S	54°38'W	24.4	24.1	24.8	24.9	24.6	24.1	24.0	24.4	25.7	24.9	24.6	24.8	24.7
C.da Araguaia-Pa	8°15'S	49°17'W	24.8	24.5	24.8	25.2	25.3	24.7	24.4	25.5	26.3	25.6	25.3	24.9	25.1
Itaituba-Pa	4°17'S	55°59'W	26.2	26.1	26.0	26.4	26.4	26.6	26.5	26.8	27.2	27.2	27.2	26.4	26.6
Marabá-Pa	5°21'S	49°07'W	25.9	25.6	25.8	26.4	26.9	26.4	26.8	26.6	26.9	27.1	26.9	26.1	26.4
Obidos-Pa	1°55'S	55°31'W	26.2	25.9	25.8	26.3	25.8	25.9	26.0	26.9	27.0	28.0	27.8	27.2	26.5
Porto de Moz-Pa	1°45'S	52°14'W	25.4	25.0	25.0	25.2	25.2	25.2	25.2	25.1	25.3	25.6	26.4	26.0	25.4
Salinópolis-Pa	0°37'S	47°20'W	26.9	26.4	25.9	25.2	26.1	26.8	26.9	27.1	27.4	27.6	27.6	27.7	26.9
Santarém-Pa	2°26'S	54°42'W	25.8	25.5	25.5	26.0	25.6	25.4	25.4	26.2	26.7	27.0	26.9	26.5	26.0
Soure-Pa	0°44'S	48°31'W	26.9	26.2	25.9	25.6	26.7	26.8	26.8	27.2	27.6	27.9	28.0	27.7	27.0
Tomé-Açú-Pa	2°25'S	48°09'W	28.2	28.1	28.0	26.2	27.7	27.6	27.5	27.7	27.8	28.2	28.2	28.4	27.9
Tracuateua-Pa	1°15'S	46°54'W	25.2	24.9	24.5	27.9	24.6	24.6	24.4	24.6	25.0	25.3	25.6	25.7	24.9
Barcelos-Am	0°58'S	62°57'W	26.1	26.2	26.3	25.8	25.6	25.5	25.4	26.0	26.0	26.4	26.5	25.6	26.0
B.Constant-Am	4°22'S	70°02'W	25.8	25.8	25.9	25.8	25.6	25.2	25.1	25.8	26.0	26.1	26.2	26.0	25.8
Carauari-Am	4°52'S	66°54'W	26.3	26.1	26.4	26.2	25.8	25.6	25.3	26.2	26.6	26.6	26.6	26.6	26.2
Coari-Am	4°05'S	63°08'W	25.2	25.2	25.4	26.2	25.3	25.3	25.4	26.0	26.0	25.9	25.9	25.6	25.5
Erunepê-Am	6°40'S	69°52'W	26.3	26.2	26.0	26.2	26.0	25.7	25.6	26.0	26.6	26.8	26.8	26.7	26.3
Fonte Boa-Am	2°32'S	66°01'W	24.8	24.9	24.9	24.8	24.7	24.5	24.3	24.9	25.2	25.3	25.3	25.2	24.9
Humaitá-Am	7°31'S	63°02'W	25.2	25.3	25.4	25.4	25.5	25.2	25.2	26.4	26.3	26.3	26.0	25.7	25.7
Itacoatiara-Am	3°08'S	58°25'W	26.7	26.4	26.4	26.5	26.7	26.7	26.8	27.8	28.1	28.2	28.1	27.6	27.1
Iauareté-Am	0°36'N	69°12'W	25.2	25.2	25.3	25.1	24.9	24.4	24.1	24.5	25.1	25.3	25.5	25.3	25.0
Manaus-Am	3°08'S	60°01'W	25.9	25.8	25.8	25.8	26.4	26.6	26.9	27.5	27.9	27.7	27.3	26.7	26.7
Manicoré-Am	5°49'S	61°19'W	26.2	25.8	26.1	25.8	26.3	26.3	26.1	27.0	27.0	27.2	27.2	26.9	26.5

Table 4 (cont.)

Station	Lat.	Long.	Jan.	Feb.	Mar.	Apr.	May	June	July	Aug.	Sept.	Oct.	Nov.	Dec.	Average
Maués-Am	3°24'S	57°42'W	26.1	26.0	28.8	26.2	25.8	26.0	26.1	26.8	26.4	27.2	27.4	27.0	26.3
Parintins-Am	2°36'S	56°44'W	27.0	26.6	26.8	25.4	27.0	27.0	27.2	28.2	28.8	29.0	28.3	27.7	27.5
S.P. Olivença-Am	3°27'S	68°48'W	27.8	25.8	25.8	26.8	25.5	25.4	25.2	25.0	26.3	26.2	26.2	26.2	25.8
Taracuná-Am	8°10'S	70°46'W	25.2	25.3	25.3	25.8	24.9	24.5	24.1	24.4	25.3	25.4	25.4	25.2	25.0
Tapuruquara-Am	0°24'S	65°02'W	26.8	26.6	26.8	25.2	25.8	25.6	25.6	25.0	26.6	26.9	27.0	26.9	26.4
Cruzeiro do Sul-Ac	7°38'S	72°36'W	24.4	24.6	24.4	26.2	24.1	23.1	22.9	25.8	24.5	4.6	24.7	24.6	24.2
S. Madureira-Ac	9°04'S	68°40'W	25.2	25.3	25.2	25.0	24.3	23.5	23.0	24.1	25.3	25.3	25.5	25.5	24.8
Amapá-Ap	2°03'N	50°48'W	26.2	25.9	26.1	26.1	26.4	26.2	26.8	27.0	27.4	27.7	27.5	27.0	26.7
Macapá-Ap	0°02'N	51°03'W	26.8	26.4	26.1	26.3	26.8	26.7	27.5	29.3	8.3	28.3	28.0	27.3	27.3
Clevelândia-Ap	3°49'N	51°52'W	24.3	24.2	24.4	24.5	24.5	24.0	24.6	25.0	25.5	25.6	25.4	24.8	24.8
Porto Velho-Ro	8°46'S	63°54'W	25.1	25.2	25.3	25.3	25.3	25.1	25.0	26.4	26.6	26.1	25.8	25.4	25.6
Boa Vista-Rr	2°49'S	60°40'W	27.7	28.0	28.3	28.2	27.0	26.2	26.1	26.6	28.1	28.8	28.6	28.3	27.7
Carolina-Ma	7°20'S	47°28'W	25.6	25.6	25.8	26.1	26.4	26.2	26.1	26.6	28.1	28.8	28.6	28.3	27.6
Grajaú-Ma	5°49'S	46°08'W	25.4	25.3	25.5	25.6	25.4	25.0	24.9	25.8	27.2	27.8	26.5	25.9	25.8
Imperatriz-Ma	5°32'S	47°29'W	25.2	25.1	25.2	25.1	25.4	24.8	24.5	25.3	26.3	26.4	26.1	25.6	25.4
São Bento-Ma	5°14'S	36°02'W	26.4	26.1	26.1	26.2	26.4	26.8	26.1	26.4	26.6	26.9	27.0	27.0	26.4
São Luiz-Ma	2°31'S	46°16'W	26.8	26.4	26.3	26.3	26.3	26.4	26.2	26.6	27.0	27.2	27.3	27.2	26.7
Turiaçú-Ma	1°41'S	45°21'W	27.0	26.4	26.1	26.1	26.3	26.2	26.1	26.6	27.0	27.3	27.5	27.5	26.7
P. Nacional-Go	10°42'S	48°25'W	25.3	25.3	25.4	26.0	25.8	24.8	24.8	26.4	27.9	27.0	25.9	25.5	25.8
Taguatinga-Go	12°25'S	46°20'W	23.8	23.7	23.8	24.3	23.9	23.2	22.9	24.4	26.1	25.7	24.2	23.5	24.1
Cáceres-Mt	16°04'S	57°41'W	26.4	26.4	26.2	25.3	23.5	22.1	21.5	23.9	26.1	26.8	26.6	26.6	25.1
Cuiabá-Mt	15°35'S	56°05'W	26.5	26.5	26.2	25.5	24.3	23.2	22.8	25.0	27.0	27.2	26.8	26.6	25.6

INEMET (1969).

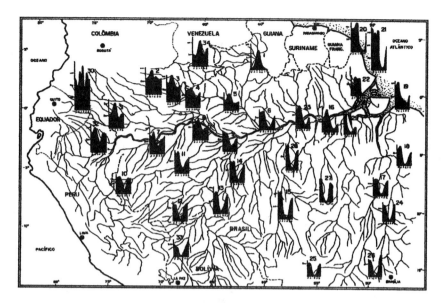

Fig. 4   **Rain distribution on the Amazonic basin in relation to the following stations: (1) Boa Vista; (2) Iauretê; (3) Tarauacá; (4) Uaupês; (5) Barcelos; (6) Manaus; (7) Benjamin Constant; (8) Fonte Boa; (9) Coari; (10) Cruzeiro do Sul; (11) Caruari; (12) Rio Branco; (13) Porto Velho; (14) Humaitá; (15) Alto Tapajós; (16) Taperinha; (17) Conceição do Araguaia; (18) Imperatriz; (19) Belém; (20) Clevelândia; (21) Amapá; (22) Macapá; (23) Parintins; (24) Porto Nacional; (25) Cuiabá; (26) Pirenópolis; (27) Serra do Cachimbo; (28) Jacareacanga; (29) Altamira; (30) Tema; (31) Average of the region; (32) Iquitos; (33) Apolo; (34) Average of the region (Salati et al., 1983)**

DNAEE in 1992. Based on this information, the Amazon basin takes in a total of 15.04×10^{12} m³ of water per year, discharges to the ocean a total of 6.59×10^{12} m³ of water per year and returns to the atmosphere, through the process of evapotranspiration, a total of 8.45×10^{12} m³ of water per year. The importance of the water vapour generated by evapotranspiration is absolutely relevant for the dynamic process, leading to the formation of clouds and originating the precipitation of the whole region. Research was developed during the 1970s and 1980s with the objective of determining the importance of this large vapour mass on the process of cloud formation and precipitation on the Amazon region. There were three lines of investigation (Salati et al., 1984). The first one was directed to the establishment of the water balance for the Amazon basin as a whole as well as for other basins of importance. The second one was to determine the water vapour fluxes, using the available information of radio wave analysis.

17

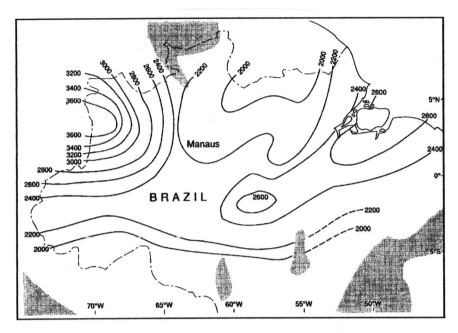

Fig. 5 **Normal precipitation in the Amazon basin in mm/yr (Salati et al., 1978)**

The third line aimed to determine the spatial distribution of oxygen[18] and deuterium isotopes in the water. The combination of these studies provided the following conclusions:

(a) the primary origin of the water vapour that penetrates the Amazon river is the Atlantic ocean, as shown in figures 6, 7, 8, and 9, which indicate the flux of the water vapour;

(b) the vapour flux is higher in the region of lower longitude, diminishing towards the west. The precipitable water, however, increases from the east towards the west. The total water vapour from the ocean is of the same order of magnitude as the water vapour produced by evapotranspiration from the forest of the Amazon basin; and

(c) the variation of the spatial distribution of oxygen[18] concentrations in rain water is lower than expected assuming the precipitation had had the Atlantic Ocean as a single source of water vapour.

As a consequence of the above observations, a model for water vapour recirculation has been proposed. According to the model, the rain from a specific region originates from clouds formed by a mixture of primary water vapour coming from the ocean, and also from

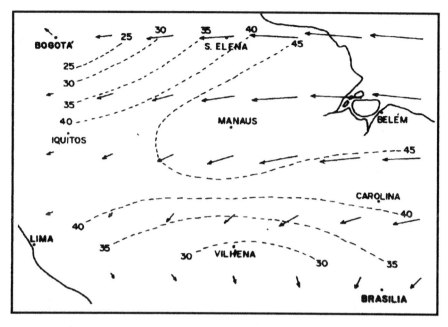

Fig. 6 **Values of vectorial field $\vec{Q} = \vec{Q}_\lambda + \vec{Q}_\lambda$. Mean of period 1972–1975, March, obtained for the 5° latitude × 5° longitude squares (1 cm = 2000 g$_v$/cm). Broken line: precipitable water in mm (Marques et al., 1979)**

the water vapour from plant transpiration. The model has allowed, in the first instance, for a better explanation of rain distribution and the observed spatial concentrations of oxygen.[18] The main consequence of these conclusions is that the dynamic balance of the atmosphere above the Amazon region is dependent on its green coverage, that is, on its forest.

This conclusion indicates that the forest is not just a simple consequence of the climate but that the present climatic conditions are dependent on the forest itself. From this point of view, the geomorphological characteristics, as well as the geographical situation, allowing the interception of moist and humid winds into the general circulation of the atmosphere, and the intertropical convergence, lead to factors determining the establishment of a hot and humid climate, which enables the development of an equatorial forest. The development of the Amazonian ecosystem through time and its present state of equilibrium, has produced the water balance as it is known today. The primary source of water vapour is the Atlantic ocean. However, studies of the water vapour dynamics of the region have

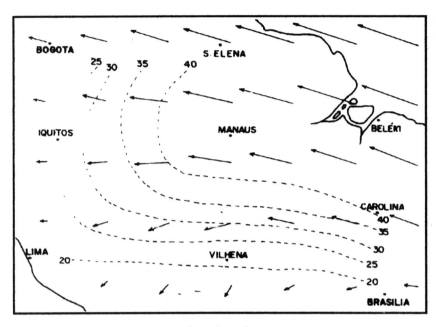

Fig. 7 **Values of vectorial field $\vec{Q} = \vec{Q}_\lambda + \vec{Q}_\lambda$. Mean of period 1972–1975, September, obtained for the 5° latitude × 5° longitude squares (1 cm = 2000 g_v/cm). Broken line: precipitable water in mm (Marques et al., 1979)**

indicated that only about 50 per cent of the present precipitation originates from this primary source of water vapour. The indigenous plants which have developed under the initial conditions of the evolving ecosystem, are today the basic constituents and, as such, basic elements for the established equilibrium, supplying through evapotranspiration the other 50 per cent of water vapour needed to produce the present level of rainfall. This maturing of the Amazonian ecosystem took place through several stages of equilibria and of selective and progressive experiments, in parallel with a constant and dynamic interaction between the biosphere and the atmosphere.

This leads to the conclusion that a large-scale deforestation would affect not only the biosphere but also the prevailing climatic conditions, including the water balance of the Amazon region and its surrounding area. Based on this information, Salati (1983) reached the following conclusions:
(a) Overall, deforestation will change the volume of water along the basin, increasing river flows during the rainy seasons while the diminishing volumes of groundwater reservoirs will reduce river

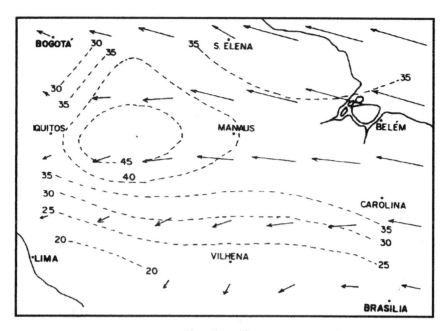

Fig. 8 **Values of vectorial field $\vec{Q} = \vec{Q}_\lambda + \vec{Q}_\lambda$. Mean of period 1972–1975, June, obtained for the 5° latitude × 5° longitude squares (1 cm = 2000 g$_v$/cm). Broken line: precipitable water in mm (Marques et al., 1979)**

flows during dry periods. A reduction of the forest area will imply less availability of water vapour in the atmosphere and, will lead to a reduced rainfall, particularly during dry periods. Climate change may take place, causing a long, characteristic dry period, with a water deficit in the soil and a broader variation of temperature;

(b) since the Amazon region is a source of water vapour for surrounding regions, large-scale deforestation will cause a reduction of the water vapour which is the source of rain in the central region of South America. Therefore, deforestation would bring about a reduction of the potential hydroelectric power available to some Brazilian regions;

(c) the solar energy reaching the region at tree top level is about 420 calories per cm^2 per day. It is estimated that 50–60 per cent of this energy is utilized for the evapotranspiration process within the forest system. In this way, deforestation changes the energy balance. A large portion of the energy utilized by the plants to stimulate water evaporation, either by transpiration or by direct

21

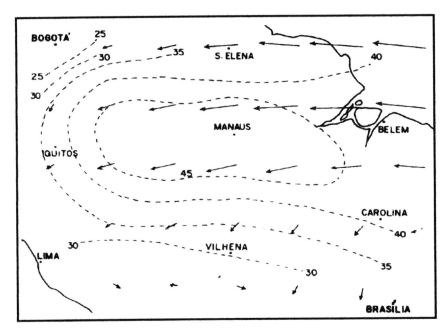

Fig. 9   **Values of vectorial field $\vec{Q} = \vec{Q}_\lambda + \vec{Q}_\lambda$. Mean of period 1972–1975, December, obtained for the 5° latitude × 5° longitude squares (1 cm = 2000 g_v/cm). Broken line: precipitable water in mm (Marques et al., 1979)**

evaporation of the intercepted water, would instead be used for warming up the surrounding air.

On the other hand, there are parts of the Amazon region where different results can be observed between areas covered with pastures and areas covered with forests. An example would be Marajó island, where forested land shows a better distribution of precipitation throughout the year, with a monthly minimum close to 80 mm. In pasture areas, the precipitation is zero during dry periods. The total rainfall of the two areas is practically the same. It has been observed that the daily temperature variations are larger on pasture land, indicating a smaller availability of atmospheric water. Observation of the clouds has shown that the locally formed cumulus are smaller, and in higher altitudes of pastures area, rather than in areas covered with forest.

Those conclusions published by Salati (1983) have been confirmed by several articles in *Amazonian Deforestation and Climate*, edited by Gash et al., in 1996. These studies have indicated an air temperature increase close to the ground (Souza et al., 1996; Hodnett et al., 1996).

The model by Lean et al. (1996), concludes that large-scale defor-estation, that is, a total forest clearing, leads to climatic modifications towards the reduction of precipitation in the Amazon area under consideration. The model forecast for total deforestation indicates a reduction of 7 per cent on precipitation, a reduction of 19 per cent on evapotranspiration and an increase of 2.3°C on ground temperature.

Preservation of the biodiversity

Considering that the Amazon region has one of the largest diversity indices of the planet, the Brazilian government has made, during the last few decades, a great effort towards the implementation of conservation units. Presently, the Brazilian Amazon contains 112 conservation units, covering an area of 43,496,837 hectares, corre-sponding to 8.7 per cent of its territory.

Table 5 indicates the number and the corresponding areas of the conservation units, according to Ryland (1995), for the different types of conservation units established by the Brazilian law. In table 6 are listed the national parks, the biological reserves and the ecological reserves of the Brazilian Amazon. The location of the stations listed on these two tables is shown in figures 10 and 11.

Water resources development

As previously explained, the Amazon region is rich in water and other natural resources. Of the 6.1 million km^2 of the Amazon basin, 3.85 million km^2 are located in the territory of Brazil. This area rep-resents 45 per cent of the Brazilian territory, estimated at 8.5 million km^2. It is then quite reasonable to assume that these natural resour-ces will be developed by the Brazilian authorities in the near future. Next to water supply for domestic purposes, the most important water uses in this basin are hydropower and navigation. Flood con-trol, from a practical standpoint, can only be approached from a non-structural viewpoint (flood warning, flood insurance etc.). Water quality in reservoirs created by hydropower development is an issue of great importance for appropriate environmental management. In this section we examine the utilization of Amazonian water resources for hydropower and navigation and the associated water quality issues in rivers and reservoirs. For reasons mentioned above the dis-cussion will concentrate on the Brazilian Amazonia.

Table 5 **Conservation units**

Category of conservation unit	Number of units	Area (ha)
Direct use (sustainable management)		
Federal		
National forest	24	12,527,986
Federal reserve for extraction	08	2,199,311
Federal area for environmental protection	02	82,600
Area of relevant geological interest	02	18,288
State		
State forest	11	1,401,638
State reserve for extraction	03	1,438,978
State area for environmental protection	10	6,922,257
Subtotal	60	24,591,058
Indirect use (integral protection)		
Federal		
National park	10	8,301,113
Biological reserve	08	2,902,800
Ecological station	11	2,007,666
Ecological reserve	03	457,574
State		
Park[1]	10	3,880,953
Biological reserve[2]	03	105,878
Ecological station	03	1,244,678
Ecological reserve	01	3,000
Subtotal	49	18,903,662
Complementary category (RPPN)	03	2,117
Subtotal	03	2,117
Total	112	43,496,837

[1] The state park of Serra do Araçá, Am, is mainly covered by the National Forest of the Amazon, Am.
[2] The state biological reserve of Morro de Seis Lagos, Am, is contained by the limits of the National Park of Pico da Neblina, Am.
Note: The total area of the conservation units which form the Brazilian Amazon, taking into account the overlap of the above referred areas, is of 41,641,237 ha.

Hydropower

Hydropower in Brazil is managed and controlled by Eletrobrás, a government agency, responsible for the planning and operation of

Table 6   **National parks, biological reserves, and ecological reserves of the Brazilian Amazon**

State	Units	Date of decree	Area (ha)
National parks (10)			
Tocantins	Araguaia	1959	562,312
Pará	Amazônia	1974	994,000
Rondônia	Pacaás Novos	1979	764,801
Amazonas	Pico da Neblina	1979	2,200,000
Amapá	Cabo Orange	1980	619,000
Amazonas	Jaú	1980	2,272,000
Mato Grosso	Pantanal Matogrossense	1981	135,000
Acre	Serra do Divisor	1989	605,000
Roraima	Monte Roraima	1989	116,000
Mato Grosso	Chapada dos Guimarães	1989	33,000
Subtotal			8,301,113
Biological reserves (8)			
Pará	Rio Trombetas	1979	385,000
Rondônia	Jarú	1979	268,150
Amapá	Lago Piratuba	1980	357,000
Amazonas	Abusari	1982	288,000
Rondônia	Guaporé	1982	600,000
Maranhão	Gurupi	1988	341,650
Pará	Tapirapé	1989	103,000
Amazonas	Uatumã	1990	560,000
Subtotal			2,902,800
Ecological reserves (3)			
Amazonas	Sauim-Castanheiras	1982	109
Amazonas	Jutai-Solimões	1983	284,285
Amazonas	Juami-Japurá	1983	173,180
Subtotal			457,574
Total			13,669,153

electrical generating, transmission, and distribution systems. According to the Brazilian Constitution, it is a prerogative of the Federal Government to explore, directly or by concession, authorization or permission for exploitating the hydropower potential of river courses in co-operation with the states where the those potential sites are located. Eletrobrás operates all over Brazilian territory through the different power-generating companies under its control. In figure 12 the different concessionaires in various parts of the country are shown. The whole Amazonian basin is operated by Eletronorte.

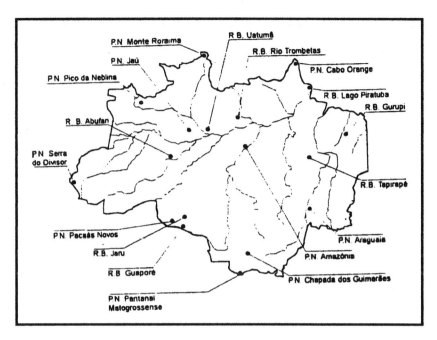

Fig. 10   **National parks and biological reserves of the Brazilian Amazon**

Although Eletrobrás operates in a decentralized way, its long and medium-range planning and operation is performed through a commission under the leadership of Eletrobrás with the participation of all the subsidiaries. The most recent plan of the electrical sector in Brazil is the so-called plan 2015 with a planning horizon for the year 2015. This plan is based on four different demand scenarios in which the GNP of the country would vary at rates ranging from 2 to 6 per cent per year from 1992 till 2015. As a result of these scenarios the forecast electrical energy consumption in Brazil is depicted in table 7. Electrical energy represents nearly 40 per cent of the total energy consumption in the country.

The electricity generating network in Brazil is predominantly hydro. The 228 TWh in 1992 had a contribution from hydroelectricity of 96 per cent. Thermal generation is utilized for isolated systems and in a complementary way to improve the reliability of the hydroelectric power system. There are three segments in the Brazilian hydroelectric system: the integrated system south/south-east/centre-west, the integrated system north/north-east and the isolated systems

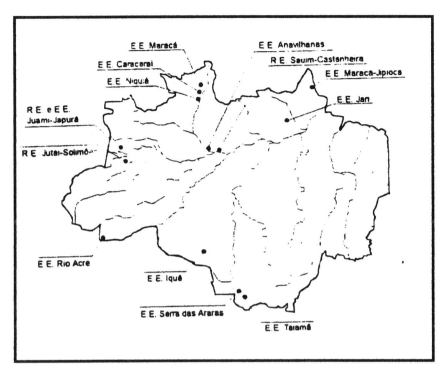

Fig. 11 **Ecological stations and national ecological reserves of the Brazilian Amazon**

of the northern region. The main characteristics of these generating and transmission systems are: reservoirs with multi-year regulation, large distances between the generating plants and the demand centres, watersheds with hydrologic diversity, a high degree of electrical integration among distinct subsystems of different watersheds and a large potential of hydropower development, notably in the Amazon basin.

Electric sources for Brazil for the planning horizon of year 2015 are depicted in table 8. It can be observed that almost the only alternative for the country for the next 20 years is hydropower. Hence, Eletrobrás will proceed to develop the potential sites for several reasons including the large potential available at lower cost when compared with other feasible options (only one quarter of this potential is developed or under construction); a renewable resource whose costs of operation do not depend on oscillations of fuel costs; available expertise in the country with respect to planning, designing and construction of hydropower plants; hydropower reservoirs can and should

27

Fig. 12 **Geographic distribution of electrical energy concessionaires in Brazil**

be used for multiple uses of water (irrigation, navigation, water supply, etc.) improving the national economy in other sectors and finally the availability of expertise in the country for transmission of electricity at large distances which would allow the development of the potential available in the Amazon.

A more detailed analysis of the hydropower potential of Brazil (table 9) shows that more than 50 per cent of that potential is located in the Amazon basin, in particular in the state of Pará. The development of the Tocantins river will have the highest priority, followed by the Xingu river basin. The priority of the upper Xingu is to supply the south-east/south/centre-west regions while the lower Xingu should supply the north east region. The capacity of electrical interconnection between the Amazon basin and the north and north-east region

Table 7   **Forecast of GNP and energy consumption increase in Brazil**

	1992	2000	2005	2010	2015
Scenario 1					
GNP (10^9 US$)	321.2	382.5	488.2	593.9	722.4
(%)	–	2.2	5.0	4.0	4.0
Energy (TWh)	224.3	293.8	384.0	467.2	563.0
(%)	–	3.4	5.5	4.0	3.8
Scenario 2					
GNP (10^9 US$)	321.2	450.9	575.5	700.2	851.6
(%)	–	4.3	5.0	4.0	4.0
Energy (TWh)	224.3	329.5	430.6	523.9	631.3
(%)	–	4.9	5.5	4.0	3.8
Scenario 3					
GNP (10^9 US$)	321.2	516.0	690.6	881.3	1,124.5
(%)	–	6.1	6.0	5.0	5.0
Energy (TWh)	224.3	360.7	473.2	589.7	731.4
(%)	–	6.1	5.6	4.5	4.4
Scenario 4					
GNP (10^9 US$)	321.2	540.8	723.7	968.5	1,295.7
(%)	–	6.7	6.0	6.0	6.0
Energy (TWh)	224.3	377.6	495.4	642.6	826.4
(%)	–	6.7	5.6	5.6	5.2

Eletrobrás (1994).

Table 8   **Energy resources for hydroelectric generation in Brazil**

Source	Potential		Cost
	Gw year	GW	(US$/MWh)
Hydro	123.5	247.0	33% < 40
			29% between 40 and 70
			28% > 70
Coal	12.0	18.0	50 to 70
Nuclear	15.0	25.0	60 to 70
Total	150.5	290.0	

Eletrobras (1996).

is higher than 5,000 MW and the south-east/centre-west varies between 3,000 to 6,000 MW depending on the energy demand scenarios. Including the Madeira and Tapajós river basins there is an additional 11,000 MW. Four hydropower plants (figure 13) would provide

Table 9   **Brazilian hydroelectric potential as firm energy (MW-year)**

Basin	Operating and construction	Survey/feasibility studies/design	Estimated	Total
Amazon	3,707.0	26,173.5	37,173.5	68,623.2
Atlantic N-NE	140.0	94.6	1,329.0	1,563.6
São Francisco	5,707.0	2,673.0	1,270.5	9,650.5
Atlantic E	909.7	5,579.9	1,327.0	7,816.6
Paraná	18,715.2	6,045.8	5,426.1	30,187.1
Uruguay	141.8	6,268.0	1,355.4	7,765.1
Atlantic SE	743.8	765.1	1,931.0	3,439.9
Total	30,064.4	47,619.7	51,361.9	129,046.0

Electrobrás (1996).

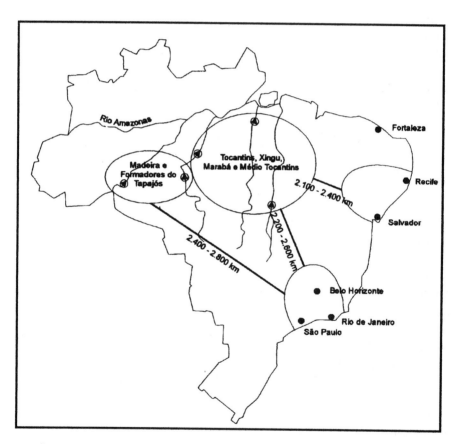

Fig. 13 **Transmission of electrical energy from the Amazon basin to different regions in Brazil**

this energy: Belo Monte (11,000 MW) and Altamira (5,720) in the Xingu river basin; TA-1 (9,528 MW) in the Tapajós river basin and MR-1 (6,854 MW) in the Madeira river basin.

The installed hydropower capacity in the Amazon basin in 1996 was 4,734 MW (Tucuruí, 4,240; Curuá-Una, 30; Coaracy Nunes, 40; Balbina, 250 and Samuel, 174). Under the Eletrobrás 2015 plan a considerable amount of hydropower is to be installed in a highly environmentally sensitive region. Consequently, a very detailed multi-objective analysis is essential in siting the associated dams and reservoirs. All related variables, economic, social, political, and environmental are taken into account at the very first stage of the implementation of the system.

The utilization of simple indexes should be avoided (Goodland, 1996) such as the ratio of inundated area per MW installed or the number of people involuntarily resettled per MW installed, since they do not encompass multiple uses of resources and other economic aspects. Concepts such as regional insertion of the project in the local community and multiple uses of water resources should play a substantial role in defining the best site for the power plants and reservoirs.

In order to illustrate the importance of the hydropower potential for the Amazon the horizons of extinction of competitive hydropower in Brazil for different development scenarios are presented in figure 14.

It can be observed that, without the Amazon potential, the extinction of the hydropower potential of the country takes place in the period 2003–2012, while if one considers the Amazon region this extinction takes place in the period 2012–2021 depending on the scenario adopted. The hypothesis of not using Amazonian hydropower implies the implementation of a significant thermoelectric programme for the country starting around 2005–2010 depending on the demand scenario. This programme would very likely rely on coal and nuclear plants. This would certainly result in higher energy costs to the final consumers and severe environmental problems related to air pollution at local and global scale and disposal of nuclear wastes. It thus becomes apparent that the adequate planning of the Amazonian hydropower plants, including economic, social, and environmental variables, is the only feasible alternative for the long-range supply of electrical energy in Brazil.

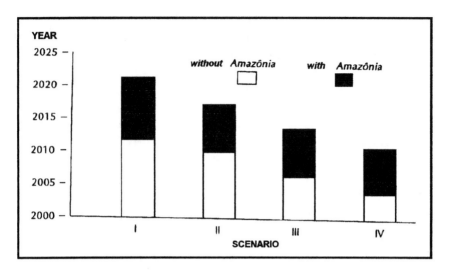

Fig. 14 **Extinction horizons of competitive and environmentally feasible hydro-power in Brazil**

Navigation

Brazil contains the largest river network in the world. Approximately 40,000 km of rivers are natural waterways for navigation. Unfortunately, due to its topography, most of the developed areas of the country are located near the coast, a region that presents great difficulties for navigation. Only 10 per cent of the land where development has taken place is located near those navigable watercourses (Cabral, 1996). Today, the migration of farmers and miners to the hinterland is allowing the use of part of this huge river network for exporting grains, minerals, oil, and construction materials and importing equipment with a load tonnage of more than 12 million tons a year. In the Amazon region large trucks of up to 45 tons of load use the roll-on roll-off system to go from Porto Velho to Manaus through the Madeira river or from Manaus to Belém through the Amazon river. Approximately 50 per cent of the 40,000 km of navigable waterways in Brazil are located in the Amazon region.

For decades fluvial navigation in Brazil has waited for a national plan that would integrate this transportation system with the general transportation system of the country, one which would compare favourably with alternative transportation systems. In 1989 Brazil's Ministry of Transportation elaborated the PNVNI – "Plano Nacional

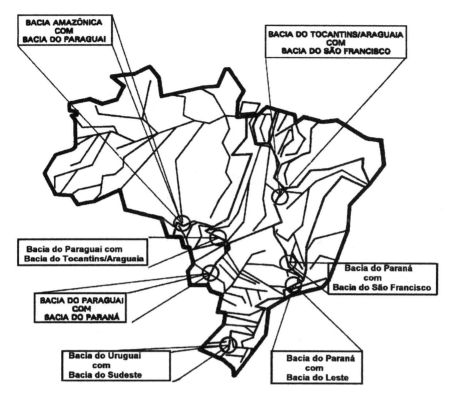

Fig. 15   **Interconnection of waterways in Brazil**

de Vias Navegáveis Interiores." This plan, with a horizon of 2002, has established load fluxes, a fluvial network and fleet, integrated with the hydropower and irrigation sectors and with planned legislation for interior navigation. Although the plan has no details of water-ways interconnection, a preliminary analysis indicates that it would be possible to go from Belém to the Plata estuary through a trans-continental waterway of 8,000 km (figure 15). Unlike the Eletrobrás 2015 plan the navigation plan has not taken off and is being subjected to major revision by the government.

A more detailed view of the waterway network of the Amazon basin including harbours and cities is given in figure 16. The Amazonas-Solimões is a river that allows all-year-long navigation of oceanic ships from its mouth in the Atlantic up to Iquitos in Peru. The Madeira river is integrated in the highway system in Porto Velho and constitutes an important axis in the export of agriculture products from the centre west farms to Manaus. The other important water-

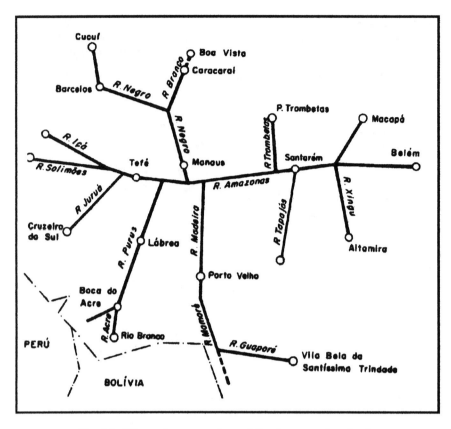

Fig. 16   **The waterway system of the Amazon river basin**

ways are the Negro, in whose riverbanks Manaus is located, the Branco, a tributary of the Negro, the Purus, and the Juruá. A major system in the Amazon dominion is the Tocantins-Araguaia (figure 17). This system crosses centre west Brazil, the new agriculture frontier of the country. It extends from its mouth in the Pará river up to the central highlands of Brasília nearly 3,000 km away. This waterway includes 715 km in the Tocantins, 1,701 km in the Araguaia and 425 km in the Mortes river and can be readily integrated with the harbour complex of Belém and the railway system of Carajás and the RFFSA (federal railway system of Brazil). The implementation of this waterway system depends on the construction of the locks in the Tucuruí hydropower plant to allow the transposition of 72 m of depth difference in the Tocantins river. In the Araguaia river the obstacle is the rapids of Santa Isabel which will be inundated by the lake to be cre-

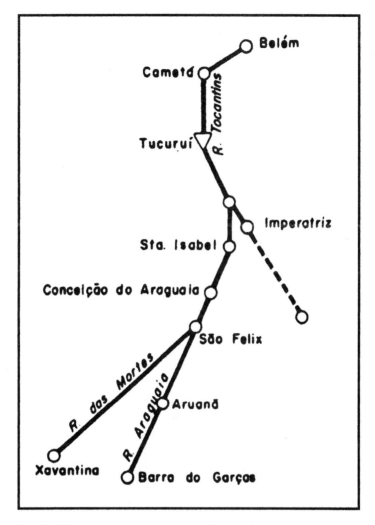

Fig. 17   **The waterway system of the Tocantins and Araguaia rivers**

ated by the hydropower plant planned for this location. At Santa Isabel dam there will be locks to transpose 60 m of difference in water levels.

The basic Amazonian waterway system encompasses in particular the Içá, Solimões, Amazonas, Juruá, Acre, Purus, Madeira, Tapajós, Xingu, Tocantins, Araguaia, Japurá, and Negro rivers. These rivers have contact points with the highway system in the Amazonia, allowing access to regions of low population density and are axes of inter-regional integration (Cabral, 1996).

35

This basic network has been under-explored for several reasons, including deficiencies in signalling and buoying systems in the several tributaries of the Amazon and Solimões, lack of appropriate harbour installations to allow the conjugation of the waterway and the high-way and the existence of rapids that make it difficult to navigate dur-ing low waters. Consequently, the solution to this problem includes the construction of bypasses and reservoirs for regulation in the Negro, Uaupés, Tiquié, Içana, and Araguaia-Tocantins; construction of small harbours at the intersection of highway BR-230 and the tributaries of the right bank of the Amazon and the improvement of the existing harbour system in the main river (Tefé, Manaus, etc.).

Legal and institutional issues

Water resources and environmental issues in the Brazilian Constitution

The current Federal Constitution of Brazil, issued on 5 October 1988, emphasizes the environmental theme with a special chapter dedicated to the subject. Chapter VI, section 225 presents the National Environmental Policy which was based in the Federal Law 6938 of 31 August 1981. According to this section, all citizens have the right to an ecologically equitable environment. This is a major national asset and is essential for a healthy quality of life. The gov-ernment, with the general public, must defend and preserve it for present and future generations. According to the first paragraph, to ensure the effectiveness of this right, the federal government has some important obligations with significance to water resources, such as:
- preserve and recover essential ecological processes and provide for the ecological management of species and ecosystems
- define, in all federate units, physical space to be specially protected
- impose environmental impact studies for licensing of civil works or any potentially harmful activity
- control the production, commercialization and usage of techniques, methods, and substances that imply threat for quality of life, life itself and for the environment.

Other constitutional statements related to water resources in section 225 are:
- the exploitation of mineral resources implies the obligation to

restore the degraded environment, in compliance with the techniques suggested by the related public agency

- conduct and activities harmful to the environment will subject the infractor to criminal or administrative sanctions, independently of the obligation to mitigate the damage generated
- the Amazon forest, the Atlantic forest, the Pantanal, and the coastal zones are national assets which will be used under conditions that warrant the preservation of the environment including the use of natural resources.

Other sections of the Federal Constitution are related to the environment. Since our interest is specifically in water resources planning and management, the next items will discuss the constitutional precepts related to water.

Water as a public good in Brazil

Brazilian Civil Law (section 65), states that all goods in the territory belonging to the Union, states or municipalities are public, all other goods are private. Public goods are classified as (section 66):

 (i) people's common use goods such as rivers, lakes, seas, roads, and streets;
 (ii) special use goods, such as buildings or lots serving the federal, state, or municipal government;
(iii) national goods, that is, those goods belonging to the Union, states, or municipalities.

An important change introduced in the Federal Constitution of 1988 was the division of waters into federal and state property. Federal waters are those flowing in rivers that flow through two or more states or that divide two states. State waters are those flowing in rivers that flow solely in the state territory. In this way, municipal waters envisioned in the Water Act of 1934 no longer exist. According to this same statute (section 225) the environment will be considered a public good. Federal Law 6938 of 31 August 1981, in accordance with these precepts, considers the environment as in public ownership, defining it as: "the set of physical, chemical, and biological conditions, laws, influences and interactions that allow, hold and reign all forms of life." Among environmental resources this law includes the interior waters, surface and groundwater, the estuaries and the territorial seas (section 3).

According to the Civil Code (section 67) things in public ownership cannot be transferred to the private sector. Pompeu (1992), quoting other counsellors, states that public goods of common use are not susceptible to the right of ownership, although the tradition allows the usage of the term to designate the holder of the judiciary relationship to whom is entrusted the care and management. In this respect the public agencies are the holders and the people and the state the beneficiaries of public goods. Section 68 of the Civil Code states that the common use of public goods can be free or charged according to specific legislation at the federal or state level. Similarly, the Water Act (section 36) states that the common use of waters can be charged in accordance with laws and rules of the administrative region where they belong. Public goods can be used by the private sector by specific authorization from the holders. In this situation the user shall pay the public agency for the right of use.

Water use and water permits

According to the Federal Constitution (section 24-VI) the Union, the states, and the federal district are jointly responsible for legislation regarding forests, fisheries, nature conservation, protection of the soil and natural resources, protection of the environment and pollution control. However, the Union has the charge of legislating privately with respect to water and energy as well as to fluvial, lake, and coastal navigation. Section 22 allows the states to legislate as well (complementary) through specific legislation regulating these matters. The current constitution, however, does not allow the states to enact additional (supplementary) legislation to deal with the special situations in such a large territory as Brazil. Section 21-XII states that the federation shall explore, directly or through authorization, concession, or permission, the hydroelectric potential of the water courses. This exploitation, however, must be performed in combination with the states where the development is planned. Although the form of joint action depends on specific law, the constitutional principle allows the possibility for states to tailor any permits to their own requirements.

It is a federal duty to implement the National Water Resources Management System. Section 21-XIX demands the creation of this system by the federal government, which is also responsible for defining the criteria for issuing water permits in the country.

The Water Act is a pioneering legal instrument enacted in 1934. Many of its precepts are still valid even after the promulgation of two constitutions in the meantime. However, many precepts have not been put into practice due to the lack of specific legislation to regulate them. According to section 36, any citizen has the right to use public waters and the use of highest priority is domestic supply. In section 43 it is stated that public waters cannot be used for agriculture, industry, or hygiene without an administrative permit except in cases of insignificant usage. These permits are given for a fixed period of time, never longer than 30 years.

The national water-resources policy and management system

Complying with the statement of section 21, chapter 19 of the Federal Constitution, the Executive Office sent to the National Congress the draft of law 2249 in 1991. This draft defines the National Water Resources Policy and creates the National Water Resources Management System. In January of 1997 the president of the republic signed Law 9433 on 8 January 1997 which regulates water use in the territory of Brazil. According to this law the National Water Resources Policy (NWRP) seeks to ensure the integrated and harmonious use of water resources towards the promotion of development and social well being in Brazilian society. Section 4 presents different instruments of the NWRP such as:
- the concession of the right of use complying with criteria and priorities established in the Water Act and subsequent legislation
- water charges for water resources utilization cost sharing in multiple use water resources works
- the institution of areas for the protection of watersheds for domestic water supply.

The National Water Resources Management System proposed will have a national collegiate, basin commissions, and an executive secretary. The directives of the system as stated in section 6 impose:
- consideration of physical, hydrological, social, economic, cultural, and political peculiarities common to large countries like Brazil
- integration of federal, state, and municipal initiatives in the planning of water use adopting the watershed as the base for regional actions
- promotion of decentralization of some federal actions through the delegation to states and the federal district as long as there is explicit interest between the parties
- encouragement of technical, institutional, and financial cooperation

among water users to achieve a larger participation in construction, operation, and maintenance of hydraulic works of common interest
– stimulation of public participation in the decision-making process.

Although the above directives show a typical decentralization proposal, in other parts of the document there are contradictory rules restraining the participation of state governments as well as water users in the decision-making process. In this respect section 12 attributes to the National Collegiate, formed with representatives of the federal government, the:

– approval of the water utilization plans of federal rivers in the whole country
– approval of the classification of water courses according to priority uses
– creation of watershed commissions establishing norms and procedures for their implementation.

The water-resources law is a major advance that the water resources sector in Brazil is experiencing today. The idea of water administration at the watershed level and the charging for water use and discharge are important concepts that will certainly bring about the improvement of water management in Brazil in general and in the Amazon basin in particular.

Initiatives of the Federal Government of Brazil

In Brazil the responsibility for development and conservation of the Amazon basin is given to the Ministry of Water Resources, Environment and Legal Amazon. This new institutional arrangement should minimize past mistakes made in the process of development of the Amazon region. In addition to the fact that only about 10 per cent of the forest has been used so far the rate of deforestation decreased substantially after the elimination (in 1991) of the subsidies provided by the government for the development of the area. In addition, the Brazilian government has already established several protected areas in the region, including national parks, biological reserves, ecological stations, national forests, extractive reserves, and Indian reserves. In November 1991, the total protected area in the Brazilian Amazon was over 116 million hectares (the Yanomami and Kaiapó reserves were established after that date), which represents about 25 per cent of the region.

The Brazilian Government is currently undertaking a major study concerning the sustainable development of the Brazilian Amazon, which will involve ecological and economic zoning of the area to avoid the repetition of the mistakes of the past. Ecological and economic zoning is the most important policy instrument for territorial management through the regulation of the dynamics of land use, according to the concept of sustainability. This policy instrument aims at integrating the available scientific and technological knowledge with the social aspirations of the people of the region. The zoning is, therefore, an instrument for negotiation and adjustment among the various development proposals for the region. Three basic types of zones are being considered in the region. They are dedicated to the following uses:

– productive zones: where the use of natural resources can ensure, with the use of improved technologies, better quality of life for the population;
– critical zones: which in view of their environmental conditions require special care for their management;
– special zones of two types: those which correspond to Indian areas, extractive reserves, and conservation units; and those which correspond to sites of relevant historical or cultural interest for eco-tourism, or strategic areas (such as national boundaries).

After four years of work, the first step of the Ecological and Economical Zoning is concluding with the Environmental Diagnostic for the Region and the establishment of a computerized data bank.

The Ministry of Environment, Water Resources and the Amazon is responsible for making sure that development activities in the Brazilian Amazon Region are undertaken in line with the concept of sustainability. The National Council for the Amazon, presided over by the President of the Republic, approved in July 1995 the Integrated National Policy for the Amazon, which is the result of an extensive evaluation of past development efforts, their successes, failures, and limitations, and is an answer to current and future challenges, in our quest to achieve sustainable development. This policy takes due consideration of the paramount importance of the water resources for the Amazon region, not only because of the need for integrated management of its multiple uses, but also because water vapour generated in the region is a crucial element in the maintenance of the climatic conditions, at least at local and regional levels.

International Cooperation and TCA

At the international level the region has today an important mechanism for water resources management: the Treaty for Amazonian Cooperation (TCA). This treaty signed on 3 July 1978 in Brasília by the Governments of Bolivia, Brazil, Colombia, Ecuador, Guyana, Peru, Suriname, and Venezuela aims at the improvement of the quality of life of the Amazonian people. TCA states that while it is important to have economic development of the area it is important to preserve the natural environment. In this respect the treaty clearly proposes: "... conscious that either the socio-economic development or the environmental preservation are inherent responsibilities of each state sovereignty and that the cooperation among the contracting parties will facilitate the compliance of these responsibilities, continuing and enlarging the joint efforts that they have been making in terms of ecological conservation of the Amazon ... resolve to subscribe to the present treaty."

The cooperation foreseen in the treaty encompasses the rational use of natural resources including water, the improvement of housing and navigation, rational utilization of the flora and fauna, coordination of the health services, scientific and technological research, implementation and operation of research institutions and centres of experimental production, organization of meetings and seminars, exchange of documentation and information, increase in the rational use of human resources, promotion of local commerce, increase in tourism at national and international level. As general principles the execution of the treaty should observe the following: unanimity in order to take any action, possibilities for the contracting parties to establish bilateral and multilateral agreements as long as they are not contrary to the aims of the treaty, recognition of the need to give special attention to the initiatives of less-developed countries that imply joint action of the parties, possibility of international organizations' participation in local projects in the region, unlimited duration of the treaty and sovereign equality of the parties.

Several meetings have taken place since the signing of TCA. They involved political as well as technical issues. Several committees have been created to implement the treaty including special committees on the following: Amazonian environment (CEMAA), transportation, communication and infrastructure of Amazônia (CETICAM), Indian affairs (CEAIA), Amazonian Tourism (CETURA), science and technology (CECTA), and health in the Amazon (CESAM). The treaty is being implemented in a gradual way taking into account the present economic and financial situation experienced by most countries of the world.

The following articles of TCA are concerned with water resource:

- Article V

  Considering the important and multiple functions performed by the rivers in the Amazonian region, in connection with the economic and social development of the region, the contracting parties shall make every effort to ensure a rational use of water resources.

- Article XV

  The contracting parties shall endeavor to maintain a permanent exchange of information and cooperation among themselves and with Latin American cooperation organizations, with respect to the action areas covered by this treaty.

Since the beginning of the 1990s the TCA member countries have made a series of commitments with respect to water management of the Amazon basin including: establishment of a hydrometeorological database of the Amazon region, promotion of the exchange of researchers in hydrology and meteorology for the purpose of research enhancement, undertaking of a surface and aerological water balance, creation of a regional centre on tropical hydrology, conducting basic research on agro-meteorology, fostering the use of remote sensing, strengthening technical cooperation at all levels in the fields of hydrology and climatology, holding biannual meetings in the field of hydrology and climatology and the establishment of a basic hydrometeorological network for the Amazonian region.

When thinking about the conservation and sustainable development of the Amazon region, particularly its water resources, it is important to understand the joint responsibilities of the countries involved. In March 1989, the countries of the region met in Quito, Ecuador, and issued the Declaration of San Francisco de Quito, which among other things, established a special commission for the Amazonian environment and a special commission for Indian affairs. Furthermore, the presidents of the states party to the Treaty for Amazonian Cooperation, meeting in Manaus, Brazil, on 6 May 1989, adopted the Amazon Declaration, which stated:

We emphasize the need that the concerns expressed in the highly developed countries in relation to the conservation of the Amazon environment, be translated into measures of cooperation in the financial and technological fields. We call for the establishment of new resource flows and concessional terms to projects oriented to environmental protection in our countries,

43

including pure and applied scientific research, and object to attempts to impose conditions in the allocation of international resources for development. We expect the establishment of conditions to allow free access to scientific knowledge, clean technologies and technologies to be used in environmental protection and reject attempts made to use legitimate ecological concerns to realize commercial profits.

Impacts of modern colonization and sustainable development

Sustainable development, while being a broad concept, stating that natural resources of today are to be preserved and made available to future generations, should be analysed through several objective and quantitative criteria. In this way it has been decided to utilize the criteria of economic, ecological, and social sustainabilities for the appraisal of projects supported by national and international agencies.

- Economic sustainability is achieved when the investment output reaches levels which are compatible with the specifications of the supporting agencies (they should at least cover related interests and the amortization of capital costs).
- Ecological sustainability is attained when system productivity is maintained over time and biodiversity is preserved.
- Social sustainability is reached when the output from the investment improves quality of life and leads to a better income distribution.

The projects implemented in the Amazon region during the last few decades have not necessarily attained the criteria for sustainable development, particularly the ones associated with the expansion of agricultural and cattle-raising activities, which have been shown to be the main factors contributing to the increase of deforestation in the region. These projects have been developed in conjunction with the construction of large highways, such as the Belém-Brasília, the Transamazônica and the BR 364, connecting Cuiabá to Porto Velho. Currently presently secondary roads have also been constructed to Rio Branco-Cuiabá, leading to a multiple interaction among governmental institutions and private enterprise, interested in timber exploitation and, recently, in agricultural and cattle-raising activities.

Unfortunately, the implementation of a long-term sustainable agriculture in the Amazon region has been no more than a challenge to administrators and scientists. The equilibrium attained by the original forest is based on a strong process of nutrient recycling, where the organic matter generated (leaves and small branches) is decomposed in the soil, by a large number of microorganisms, particularly the microarthropodes. From this interaction results the breakdown of organic matter, the release of nutrients and the maintenance of the soil chemical and physiochemical properties. Deforestation interferes with this dynamic process, particularly through the destruction of thousands of living animal and vegetable species, which would, otherwise, interact within the ecological system. On the other hand, conventional agricultural activities are associated with a continuous process of planting and harvesting of a reduced number of vegetal species. The heavy rain associated with cultivation practices, devoid of soil conservation techniques, leads to soil erosion by surface runoff, or by lixiviation.

Other types of project implemented in the Amazon are related to mineral exploitation, construction of hydroelectric plants, mining of gold ore and precious stones, as well as timber exploitation, not necessarily associated with farming activities. All these activities have caused, at different levels, heavy environmental drawbacks in the region. The observed impacts are sometimes due to direct action, such as mining, particularly of gold ore, which has brought about deforestation, the destruction of river banks, the contamination of water courses by mercury, and the promotion of negative contacts with the local Indian tribes. Indirect impacts have also occurred, such as the construction of large artificial lakes for power generation, leading to uncontrolled development and to several other negative effects such as the ones caused by filling the reservoirs without previously clearing the vegetation.

It is actually very difficult to find, within the Amazon region, a neat example of a project resistant to critical analysis of the three criteria to be attained, in order that the development associated with them is considered as sustainable. Notwithstanding, this situation does not imply that such a level of development is not possible. Recent analysis (Salati, 1997), aiming at the identification of the limiting factors needed to achieve sustainable development, led to the following classification:

– natural factors
– technological factors
– educational factors
– economic factors
– institutional factors.

When developmental projects are under analysis, one or more of these factors are usually identified as limiting factors. Apparently, the main problem lies on the insistence on or initiative in implementing intensive colonization in a region where the factors or forces maintaining the dynamic equilibrium are not fully understood. Looking at the Amazon forest from above, particularly over the high density forest, it is not possible to characterize lack of nutrients, and preliminary surveys have indicated a high rate of photosynthetic primary production. This large rate of primary productivity leads to the erroneous conclusion that monocultures would have the same long-term productivity efficiency. Many decades of research have been made necessary and large areas have been deforested before having these false premises dismantled. Nowadays, support agencies do not allow the substitution of forests by parstureland, and have ruled that more than 50 per cent of the original forests should be preserved on developmental sites in the Amazon region.

While lessons have been learned over time and the legislation has established conditions for a sustainable development, the lack of institutions able to control and to enforce the law is still the most important limiting factor for ecological balance in the Amazon region.

Conclusions and recommendations

This paper has presented an overview of the Amazon basin in terms of its ecological system, the alternatives for its sustainable development and the legal and institutional framework for the development of the region from a Brazilian perspective and in the realm of the Treaty for Amazonian Cooperation. The Amazonian dominion which extents beyond the Amazon river basin has much diversity in terms of ecosystems and flow regimes. The fragility of the ecosystems should be taken into account when considering alternatives for the development of the region.

The most immediate use of the water in the basin is for hydropower and navigation. The energy matrix of Brazil shows very little resource other than hydropower. The available alternatives (coal, oil, nuclear) are more expensive and impose heavy environmental pollution. Consequently, the utilization of the Amazon basin to supply electrical energy markets in south and north east Brazil is being considered, beginning in the year 2000. Navigation is a natural transportation system in the Amazon. Today, large vessels sail from the ocean to more than 2,000 km inland. The integration of navigation with other systems (railway and highway) will be considered in the Brazilian navigation plan for the year 2000.

References

Barbosa, M.N. 1996. *Deforestation assessment and detection programs in Brazil*, Pilot programme to conserve the rainforest meeting. Bonn, Germany.

Cabral, B. 1996. *O papel da hidrovia no desenvolvimiento sustentável da região amazônica brasileira.* Brazilian Federal Senate Press, Brasilia (in Portuguese).

Comision Amazônica de Desarrollo y Medio Ambiente 1994. *Amazonia Sin Mitos.* Colombia: Editorial La Oveja Negra.

Gash, J.H., Nobre, C.A., Roberts, J.M., and Vistoria, R. 1996. *Amazonian Deforestation and Climate.* New York: John Wiley & Sons.

Hodnett, M.G. et al. 1996. "Comparison of long-term soil water storage behavior under pasture and forest in three areas of Amazônia." In: J.H. Gash et al., eds. *Amazonian Deforestation and Climate*, p. 57. New York: John Wiley & Sons.

IBGE/FBDS/FUNCATE/SAE 1995. "Diagnóstico ambiental, análises temáticas e sistema de informação geográfica." *Subsídios para o Macrozoneamento Ecológico da Amazônia Legal*, 6 vols. Rio de Janeiro.

Instituto Nacional de Meteorologia (INEMET) 1984. "Atlas Climatológico do Brasil." In: H. Sioli, ed., *The Amazon – limnology and landscape ecology of a mighty tropical river and its basin*, p. 92. Dordrecht: Dr. Junk Publishers.

Lean, J., Bunton., R., Nobre, C.A., and Rowntree, P.R. 1996. "The simulated impact of Amazonian deforestation on climate using measured ABRACOS vegetation characteristics." In: J.H.C. Gash, C.A. Nobre, and J.M. Roberts, eds. *Amazonian deforestation and climate*, p. 549. New York: John Wiley & Sons.

Marques, J., et al. 1979. "O campo do fluxo de vapor d'água atmosférico sobre a região Amazônica." *Acta Amazônica*, 9(4): 701–713.

Mattos de Lemos, H. 1990. "Amazônia: In Defense of Brazil's Sovereignty." *The Fletcher Forum of World Affairs* 14(2).

Prance, G.T., and Lovejoy, T. (eds.) 1985. *Amazônia Key Environments.* Oxford: Pergamon Press.

Repetto, R. 1988. "Economic Policy Reform for Natural Resource Conservation." *World Bank Environment Department*, Working Paper No. 4. Washington, D.C.

Ribeiro, M.N.G. et al. 1982. "Radiação solar disponível em Manaus e sua relação com duração do brilho solar." *Acta Amazônica*, 12(2): 339–346.

Salati, E. et al. 1978. "Origem e distribuição das chuvas na Amazônia." *Interciência*, 3(4): 200–206.

Salati, E., Schubart, H.O.R., Wolfgang, J., and Oliveira, A.E. 1983. *Amazônia: desenvolvimento, integração e ecologia.* São Paulo: Brasiliense, CNPq.

Salati, E., and Marques, J. 1984. "Climatology of the Amazon region." In: H. Sioli, ed. *The Amazon: limnology and landscape ecology of a mighty tropical river and its basin*, ch. 4, pp. 85–126 Dr. Junk Publishers, Dordrecht.

Salati, E. et al. 1990. *"Amazônia." The Earth as Transformed by Human Action.* New York: Cambridge University Press.

Salati, E., Santos, A., Lovejoy, T., Klabin, I. 1997. *Porquê Salvar a Floresta Amazônica.* Instituto Nacional de Pesquisas da Amazônia (INPA).

Sioli, H. 1984. *The Amazon: limnology and landscape ecology of a mighty river and its basin.* Dordrecht: Dr. Junk Publishers.

Sousa, J.R.S. et al. 1996. "Temperature and moisture profiles in soil beneath forest and pasture areas in eastern Amazônia." In: J.H.C. Gash, et al. eds. *Amazonian Deforestation and Climate*, pp. 125–137. New York: John Wiley & Sons.

2

The Amazon Policy of Colombia

Fabio Torrijos Quintero

Geographical location and biophysical features

The Amazon region, which covers 336,583 km², accounts for one-third of the national territory. It comprises the departments of Amazonas, Caquetá, Guainía, Guaviare, and Putumayo. Geographically, it extends from the Guaviare and Guayabero rivers in the north to the trapezoid stretching between the Amazon and Putumayo rivers in the south, and from the Brazilian border in the east to the watershed of the eastern cordillera in the west (map 1).

The Colombian Amazon region is marked by biophysical, cultural, and socioeconomic diversity. The many different climates, geological formations, and altitudes make for highly divergent landscapes, with a great variety of soils, plants, and biodiversity. Rainfall is abundant, ranging from 2,500 to 4,000 mm a year, and the average temperature is 25°C.

The region's natural resources have been deteriorating under the impact of settlement, which has led to the deforestation of 7,500 ha of tropical rainforest, affecting species of fauna and flora and thus entailing a loss of biodiversity; pollution of water sources by waste-water, organic residues, and the chemicals used for illegal crops; and soil degradation from inappropriate farming practices. The illegal

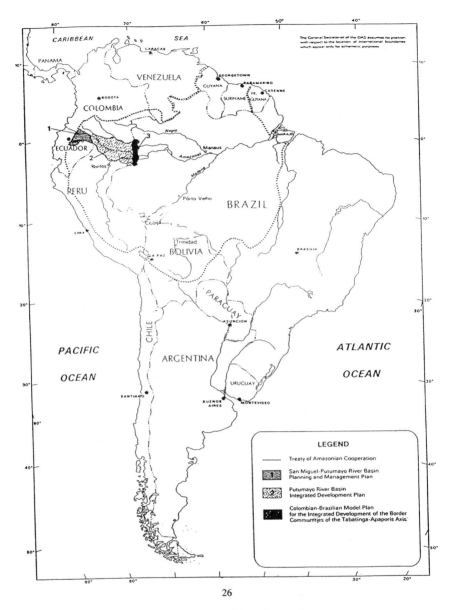

26

Map 1   **Location of frontier projects**

occupation of parks, natural reserves, and the Amazonian forest
reserve has triggered various types of conflict, which call for careful
examination to identify solutions consistent with local dynamics and
with the objectives of the protected area system. Wildlife has also suf-

fered from the consumption needs of the settlements and the illegal trade, to the point that various species are in danger of extinction.

Socio-economic features of the region

The indigenous population numbers around 60,000, and the immigrant population from several waves of colonization is 850,000. Population growth in the region has soared in the past few years, by 230,000, from 1985 to 1993. This is double the rate for the country as a whole, the highest of any region, and 45 per cent of it stems from migration. Of this population, 52.5 per cent lives in poverty and 24.3 per cent in extreme poverty.

The indigenous population is composed of 59 different ethnic groups, living primarily in rural parts of the Vaupés, Amazonas, and Guainía departments. The majority of them (70 per cent) are settled in 25 protected areas totalling about 18 million ha; the rest live in unprotected zones. The 1991 Constitution established that indigenous lands are territorial entities, and as such have the right to be governed by their own authorities, who are entitled to exercise the relevant powers, manage resources, levy any taxes needed to perform their functions, and share in the national revenue. The indigenous territories are governed by councils formed and regulated according to community practices and customs. Among other things, they enforce the regulations on community land-use and settlement within their territories and see to the preservation of their natural resources.

Illiteracy is 18.4 per cent, compared to 12.7 per cent for the nation as a whole. Primary and secondary school attendance is below the national average. Access to higher education has improved in the region, as a result of specially designed policies and the strengthening of the University of Amazonia.

The region's economy is based on highly diverse activities, as heterogeneous as its population. The economic activity of the indigenous communities is centered around small farms and the sustained use of the forests, and is basically oriented towards creating family-size units.

For indigenous communities that are highly integrated into the market community, the traditional system of production has gradually been replaced by farming geared essentially to the local consumer market. The economic activities of the non-indigenous population are based on both extraction and production. The main extractive activities at the present time are timber, oil, and gold, and to a lesser extent ornamental and food fish, in addition to wildlife.

Oil operations in the region are located in Putumayo department, and royalties constitute an important source of local revenue. However, these operations have a negative environmental impact and promote migration that affects the regional economy.

Agriculture as practised by the non-indigenous people has introduced foreign methods and products unsuited to the region's environmental conditions. No appropriate technology has been developed for livestock raising, so that there are no regional economic alternatives that are environmentally viable. Fishing, which takes place primarily in the departments of Amazonas, Caquetá, and Putumayo, is characterized by uncontrolled exploitation, and lack of scientific knowledge of the resources has led to the depletion of certain species.

Gold production in the region, though just beginning, is causing serious environmental disturbances, that are causing international and inter-ethnic conflicts. An economic model based on the cultivation and processing of coca has been developed, and dominates the region's economy. The national government's decision to eradicate illegal crops and control inputs has generated a conflict of major proportions in the social and political life of the region.

The transportation infrastructure is quite weak, which affects communication both within the region and with the outside. Given this situation, coordination among the various means of transportation is inadequate, thus making it impossible to build a multimodal system. Roads are few and poor in quality, and they tend to damage the environment because the area they traverse is so fragile. Here there is a conflict between the need to create the infrastructure required to link markets and the resulting environmental impact. This conflict is intensified by the contradictory approaches of different government institutions.

River transport, considered suitable because of its low costs and minimum environmental impact, is hampered by sedimentation of the principal rivers which reduces navigability. Air transport, the ideal means of communication, suffers from an insufficient and inadequate airport infrastructure, and its high cost is also an obstacle to its use.

National environmental policy

The national environmental policy is based on sustainable human development and has five basic objectives: (1) to promote a new development culture; (2) to improve the quality of life; (3) to promote clean production; (4) to work towards sustainable environmental management; and (5) to guide the behaviour of local communities.

The increasing environmental degradation in Colombia requires effective, dynamic government intervention, with the committed support of society and production sectors, to ensure a healthy environment for everyone and incorporate the environmental costs into development considerations. Environmental policy is put into practice through education strategies, joint efforts to increase social capital, gradualism, decentralized management of national policies, public participation, and scientific and technological support.

To move towards sustainable human development, a plan to solve the major environmental problems, to prevent the deterioration of the most strategically valuable ecosystems, and to build the foundations for a new development culture so as to bring about long-term change, is being carried out. The plan is divided into two parts: environmental improvement activities and modes of action.

Programmes and activities for environmental improvement: are (1) protection of strategic ecosystems; (2) better water; (3) clean seas and coasts; (4) more forests; (5) better cities and settlements; (6) a population and settlement policy; and (7) clean production.

Modes of action

To achieve the objectives that have been established, five modes of action have been developed: (1) environmental education and consciousness-raising; (2) institution-building, through the National Environmental System (SINA); (3) the production and democratization of information; (4) environmental and land-use planning; (5) global cooperation.

Environmental management is financed out of the national budget, external credits, technical cooperation, funds managed for other institutions, and resources of territorial agencies and regional corporations.

Development projects in the Amazon region

On 20 September 1996, Law 318 was promulgated by the Colombian Congress. Under Chapter II, Article 5 of this law, the Colombian International Cooperation Agency was established as a national government institution attached to the National Planning Department, with legal status and its own separate capital and administration. The main responsibility assigned to the Agency is the coordination, management, and promotion of all non-reimbursable international technical and financial cooperation received or granted by the country as official development aid to public agencies, together with any funds obtained through debt relief for social or environmental purposes.

In addition, Colombia has signed bilateral agreements with neighbouring countries to develop specific activities and programmes.

Colombian-Ecuadorian Agreement for Amazonian Cooperation

The Colombian-Ecuadorian Agreement for Amazonian Cooperation was signed in March 1979 in Quito. It was the pioneer bilateral cooperation agreement under the Framework Treaty for Amazonian Cooperation.

The agreement reiterates the "good neighbour" principles governing relations between adjacent countries, the traditional friendship between Colombia and Ecuador, and the preservation and rational use of natural resources in the Amazon region shared by the two countries. Special importance is assigned to improving substantially the quality of life of the inhabitants. The agreement also mentions the Putumayo Declaration of 25 February 1977 which deals with similar matters.

A Joint Colombian-Ecuadorian Committee for Amazonian Cooperation was set up as a high-level group in charge of studying and coordinating programmes of common concern. It was decided that its meetings would be chaired by an official with the rank of Ambassador.

The Joint Committee was assigned a number of tasks, including bilateral cooperation in assessing and investigating existing flora and fauna; better use of the agricultural, fishery, forest, mineral, and industrial resources of the zone; the expansion and improvement of roads and interconnections; and the establishment of a cross-border air service. Importance was attached to the creation of a regular transport service on the Putumayo and San Miguel rivers, the identification of engineering works required to make the rivers navigable, the joint coordination and provision of health and sanitation services, education, fishing, mining, and the marketing of products of local origin.

Plan on land-use and management in the San Miguel and Putumayo river basins

Under this bilateral treaty between Colombia and Ecuador, the two governments have initiated coordinated activities which are fully consistent with the "good neighbour" philosophy followed by Colombia in its dealings with adjacent countries.

In accordance with the terms of the Treaty for Amazonian Cooperation, Colombia and Ecuador, aware that integration and being a good neighbour are inseparable from and consistent with an improved standard of living for the inhabitants of the outlying Amazon region, and that all this is contingent on the harmonious, sustainable, and rational development of its resources, signed the Bilateral Agreement for Amazonian Cooperation in March 1979. It was subsequently implemented in February 1985 with the Rumichaca Declaration, in which the governments reiterated their decision to promote cooperation to further the integral development of their border areas. They also approved the terms of reference for the Plan on Land-Use and Management in the San Miguel and Putumayo River Basins.

In application of Article XVIII of the Treaty for Amazonian Cooperation, Colombia and Ecuador presented a joint request to the Organization of American States for technical cooperation and collaboration to begin preliminary studies for effective management in their border areas. In 1986 the proposal for the San Miguel–Putumayo Plan (PSP) was approved, and to assist in formulating it, the Organization created the Plurinational Project on Amazonian Cooperation.

To implement the planning, diagnosis, and evaluation stages of the project, Colombia chose the Colombian Water and Land Development Institute of the Ministry of Agriculture (HIMAT) and Ecuador the Agrarian Regionalization Department of the Ministry of Agriculture (PRONAREG). For the programme and project to formulation stage, Colombia designated the Amazonian Scientific Research Institute (SINCHI) and Ecuador the Institute for Ecodevelopment of the Ecuadorian Amazon Region (ECORAE).

The general objectives of the Plan are to develop the Amazon region by following an approach that combines management and preservation of the environment and biodiversity, while taking into account the potential and limitations of the area's natural resources and the needs and aspirations of present and future generations.

The PSP seeks to rationalize the impacts of disorganized settlement of Amazonian lands, and to help check the serious damage to the

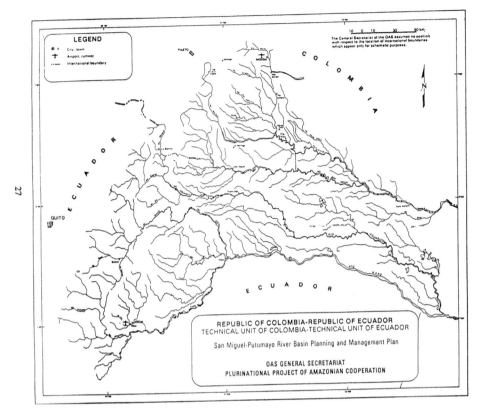

Map 2   **San Miguel–Putumayo River Basin Planning and Management Plan**

environment and natural resources caused by man's irrational use of the habitat.

The region covered by the Plan comprises 47,307 km² in the San Miguel, Putumayo, and Aguarico river basins and along the left bank of the Napo river, in Napo and Sucumbios provinces in Ecuador, and Putumayo department in Colombia (map 2).

The guidelines of the strategy for meeting the proposed objectives of the PSP are as follows:

Regional action based on an integrated approach; development of alternatives to ensure the preservation of the largest possible area of natural virgin land, including restrictions on new settlements; land-use planning and consolidation in the areas currently occupied, to guide them towards sustainable development; promotion of self-management in indigenous and settler communities; priority attention to the social needs of the most vulnerable groups; promotion of security, maintenance of peace and order, and harmonious development along both sides of the border; and strengthening of the management capacity of the agencies executing the official and private programs.

The implementation of binational activities, together with individual efforts made by the countries, will make it possible to deal with common problems and situations using a joint, planned approach, which is more effective than the traditional practice of stopping programmes at the border.

With this in view, the Plan is directing efforts toward the rational use and integrated management of natural resources, through productive projects appropriate to the ecosystem; the management of national parks and ecological reserves; restoration of the culture and traditions of indigenous communities; preservation of biodiversity; guidelines for farm production; and a programme of community consciousness-raising and training on the Amazonian environment. It also includes environmental education and research and the strengthening of community institutions as part of a training programme in which the community has the leading role in every case.

Thus, the Action Plan contains five programmes, each with its own projects, components, and activities, broken down as follows: environment; organization of production for sustainable development; attention to indigenous communities and groups; health and environmental sanitation; and community organization and training. The programmes, structured as modules, will make it possible to integrate fully the indigenous people, settlers, and the national, regional, and municipal officials involved, with self-management as a key to the success of the programmes.

The phase of pre-feasibility studies and approval of the general concept of the PSP by the two governments has been completed. The national entity in each country that will act as the executing agency is now being chosen. In Colombia it will be Corpes Amazonas, with responsibility for following up, coordinating implementation activities, monitoring projects, and supervising and managing resources. At the same time, mechanisms and procedures for obtaining external funds are being examined, to complete the financing of the various projects.

The total cost of the PSP has been estimated at US$200,870,000, of which 60.4 per cent is to be financed by the beneficiaries and 39.6

Table 1   **Costs, contributions, and resources for the San Miguel–Putumayo Plan**

Programme	Cost (US$)	Country contributions (US$m)	External resources (US$m)
Environment	23,212	4,237	18,975
Organization of production for sustainable development	143,288	116,666	26,322
Attention to indigenous communities and groups	8,357	1,847	6,550
Health and environmental sanitation	19,040	7,859	11,187
Community organization and training	4,742	725	4,017

per cent by the national governments. Of the 39.6 per cent, 15 per cent will come from the governments, and the rest from international agencies, nongovernmental organizations, or countries interested in investing in the sustainable development of the region.

The cost of each of the programmes and the amounts of the countries' contributions and the outside funds needed are presented in table 1, in millions of US dollars.

Efforts are being made to find non-reimbursable financing, to the extent possible. Such financing is currently quite limited; most international organizations use it for studies, and fund some programmes with concessionary credits and others with loans at interest rates and repayment periods that are very close to those prevailing in the financial markets. This complicates the process, since most of the projects are social in nature, with negative rates of return on capital, and cannot earn enough to attract investors. The benefit of the projects can be estimated is terms of the well-being of the community, environmental sanitation, and protection of the ecosystem and biodiversity, all of which have a significant influence on the future of mankind.

The Unit for Sustainable Development and the Environment of the OAS General Secretariat convened a meeting in Washington, D.C., in June 1996, of representatives of the countries involved and possible donors, including developed countries, international institutions, and nongovernmental organizations interested in the subject. The potential donors expressed an interest in the bilateral plans and indicated that they would look at them and consider their financial viability selectively and by country.

Colombian-Peruvian Treaty for Amazonian Cooperation

The Treaty for Amazonian Cooperation between Colombia and Peru was signed in Lima on 30 March 1979, to give greater importance to environmental preservation and the rational use of natural resources in the economic and social development of their Amazon regions.

The bilateral treaty attaches special importance to the Act Supplementary to the Protocol of Friendship and Cooperation signed in Rio de Janeiro on 24 May 1934, which had laid the groundwork for cooperation between the two countries in their Amazonian regions, and takes into consideration the Multilateral Treaty for Amazon Cooperation signed in Brasília on 3 July 1978.

Article XIV provides for the establishment of a Joint Colombian-Peruvian Committee on Amazonian Cooperation, a standing group to study and coordinate programmes of common interest to their neighbouring Amazon territories, in accordance with the treaty.

The Joint Colombian-Peruvian Committee for Amazonian Cooperation will promote assessments, research, cooperation, and joint action to expand and improve the road network and build other communications infrastructure in the border area. It will also look into the prospects for creating cross-border air routes, on protecting the ecology, and preserving the environment.

The Joint Committee is also supposed to conduct ongoing evaluations of compliance with the 1938 Agreement on Customs Cooperation, with a view to updating it and amending it to meet current needs.

Plan for the integrated development of the Putumayo river basin

With technical support from what was then the OAS Department of Regional Development and Environment under the Plurinational Project on Amazonian Cooperation, Colombia and Peru decided to draw up a binational plan for developing their common Amazonian border region in the basins of the Napo, Putumayo, Caquetá, and Amazon rivers, which has been adversely affected by unregulated settlement and by the incursion of Colombian guerrillas linked to drug trafficking and their counterparts in Peru (map 3).

The plan covers an area of 160,500 km^2. The most serious problem facing this area is the deteriorating quality of life and living conditions, which have been severely affected by the introduction of production systems and cultural and social patterns that are unsuited

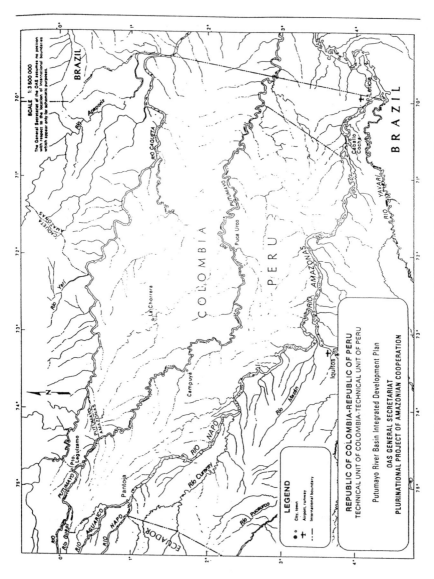

Map 3 **Putumayo River Basin Integrated Development Plan**

60

to the Amazonian ecosystem and are gradually destroying its bio-diversity and irreversibly damaging its environment.

The enormous amount of territory involved and its distance from the centres of national government has enabled settlers eager for adventure and easy riches to occupy it unencumbered by planning or development criteria. The negative effects can be summed up as reduction of the habitat of the indigenous communities, lack of basic and social services, indiscriminate cutting of tropical rainforests, indiscriminate destruction of natural resources, sedimentation, and serious contamination of rivers and soil erosion, among other things.

The situation was further complicated by the arrival of drug traf-fickers attracted to the area by the facilities it offered them. To com-pound matters, settlers and traders discovered that Amazonian timber fetched very good prices on the markets of developed countries, and so they began indiscriminately cutting down native species, most of which were irreplaceable, using methods and techniques that led to desertification and sedimentation.

The Frontier Development Plan is intended to organize the abut-ting Amazonian territories and deals with the most sensitive issues, such as immigration, colonization, introduction by white settlers of inappropriate techniques of production and habitat maintenance and of cultural systems and habits foreign to the region, all of which have severely affected indigenous ethnic groups, polluted the rivers, and given rise to an illegal trade in live species in danger of extinction.

This binational plan consists of five programmes: natural resources and ecosystems, social development and infrastructure, trade and transport, productive activities and management, and institutional organization. In these programmes 12 projects have been identified, and the pre-feasibility stage has been completed for those that follow for integral and sustainable management of forests, natural parks, environmental education, integral fishing, management of wildlife in reserves, integral services for native indigenous communities, integral health, basic sanitation, and marketing.

The benefits of the plan are unquestionable. It means that under-standing and joint management of the region and its problems, linked to the "good neighbourliness" that is an intrinsic part of Colombian foreign policy, will lead to binational activities carried out compre-hensively to meet the basic needs of the people by means of decen-tralization and institution building. This will achieve a planned use of the land that will help to drive out the evil scourge of illegal crops and trade.

The plan is in the diagnosis and pre-feasibility phase, and steps are being taken to secure financing from governments, government agencies, and nongovernmental organizations, to ensure the success of the tremendous efforts that these countries have been making, together with the OAS, in this priority field of action.

Colombian-Brazilian Agreement for Amazonian Cooperation

On 12 March 1981, the governments of Colombia and Brazil, wishing to further bilaterally the purposes and objectives of the Treaty for Amazonian Cooperation, signed the Agreement for Amazonian Cooperation under the terms of Article XVIII of that treaty, with a view to ensuring the rational development of the Amazon's resources, preserving the environment, and making use of the flora and fauna, in accordance with the principles of the Agreement for the Preservation of the Flora and Fauna of the Amazon Territories, dated 20 June 1973.

Article I of the Agreement reads as follows:

The Contracting Parties have decided to initiate a process of active cooperation to conduct joint activities and an exchange of experiences in the area of regional development and scientific and technological research as applicable to the Amazon Region, with a view to achieving the harmonious development of each of their Amazon territories, to the benefit of their people and while adequately protecting the ecology of the zone.

Special importance was attached to navigation on inland waterways and to regular passenger and cargo transport on the Amazon, Putumayo-Iça and Negro rivers, pursuant to treaties in force. Surveys and hydrographic charts of these rivers will be made and the necessary studies will be carried out to improve navigability.

Studies will be initiated in the areas of telecommunications, regular cross-border air services, road connections, and health and tropical-disease control and, in particular, measures for the proper management of natural resources.

To fulfil these objectives, the Joint Colombian-Brazilian Committee for Amazonian Cooperation was set up. It is responsible for coordinating projects established under this agreement and other programmes of common interest directed toward the harmonious development of the Amazon regions. At its first meeting, held in Leticia, Colombia, in 1987, approval was given to the Colombian-Brazilian Model Plan for the Integral Development of the Neighboring Communities along the Apaporis-Tabatinga axis.

Colombian-Brazilian Model Plan for the Integrated Development of the Neighboring Communities along the Apaporis-Tabatinga Axis

The Colombian-Brazilian border area, a major segment of the Amazon, consists of the Colombian Amazonian trapezoid, in the departments of Amazonas and Vaupés, and the state of Amazonas in north-western Brazil.

The national border on the Apaporis-Tabatinga axis, the shared waters of the Caquetá and Putumayo rivers, and the inevitable isolation from urban centres have led the communities to create neighbouring settlements, which have stamped the region with its own special imprint of mutual support and development.

It is for this reason, on the basis of the Treaty for Amazonian Cooperation, that the Model Plan of the Neighboring Communities of the Apaporis-Tabatinga axis was approved (map 4). Work began under the Plan with the technical support of the OAS, and nine projects were designed: on productive activities, social infrastructure, social development, integral services to indigenous communities, public-health infrastructure, and institutional strengthening (urban development). The initial interest in the project continued to grow, and the diagnosis phase was completed; work is now progressing on the feasibility stage and on obtaining financing.

The Plan covers a large part of the common border area, but it only regulates half of the shared line. In 1991, at a meeting of the Joint Committee for Amazonian Cooperation held in Brasília, Colombia presented to Brazil a proposal to expand it to the remainder of the border area, from Apaporis to Piedra del Cocuy, where Colombia, Brazil, and Venezuela meet.

This proposal was well received, but Brazil claimed a lack of funds for it. It is badly needed to deal with the problems originating in these areas, since this is the source of the headwaters of the Negro river, an important tributary that joins the Amazon at Manaus.

The most serious problem in the sector is the increasing and highly dangerous pollution of the rivers of the Orinoco basin, the Vaupés, and the upper Negro by mercury and other elements from inappropriate gold-mining techniques. The violent attack on natural resources must be curbed by governments. The situation in the region is further aggravated by the presence of undesirables engaged in drug trafficking, common crime, and guerrilla warfare and the near or total absence of government authorities.

Map 4   **Colombian-Brazilian Model Plan for the Integrated Development of the Neighbouring Communities along the Apaporis-Tabatinga Axis**

64

It is hoped that the parties will once again give this problem the attention it deserves so that the activities and plans agreed on in the Bilateral Treaty may be carried out. The initial results were quite encouraging and provided incentives to the people in the region. The "good neighbour" policy exerted a strong influence and, despite the slowness of the work, has stimulated local activities of cooperation and understanding, which could surely be channelled into greater achievements.

Proposed Colombian-Venezuelan Amazonian Development Plan

At the Third Meeting of Foreign Ministers, held in Quito in March 1979, the Colombian Minister of Foreign Affairs proposed to Venezuela a Treaty for Amazonian Cooperation to take bilateral action on land-use planning in the rivers of the Orinoco basins region that drain into the Amazon. The involvement and interest of international agencies and nongovernmental organizations were at a high point at the time, so it appeared highly likely that the initiative would be well received.

This also completed the Colombian government's efforts to ensure the integral binational, sustainable management of its Amazon border areas, since the agreement would cover the areas starting at Piedra del Cocuy that were not already receiving special treatment. The initiative was welcomed, but no reply was received from the Venezuelan authorities.

Conclusions and recommendations

Programmes of technical cooperation with bordering countries take on added importance when policies and strategies designed as part of a "good neighbour" approach are to be put into practice.

The national parliaments should exercise appropriate legislative control and make sure that the laws and treaties they approve are implemented. This power should be used as a natural ally of the cooperation agreements, since the benefits they offer and the problems they would solve directly concern and affect above all the poorest groups in each country.

The good-neighbour policy has fulfilled its original objectives and maintains its validity as the most important and effective means for generating mutual trust. It is important to point out that confusion as to this mechanism or the deliberate assignment to it of the execution of plans and projects, while ignoring its legal function of putting international agreements into practice, has brought the system to a standstill, to the detriment of all the parties involved.

The management and coordination of international technical cooperation should be handled by the agencies and organizations in each country in charge of executing the development plans, under the strict surveillance, and with the ongoing evaluation and coordination of the ministries of foreign affairs, in fulfilment of their constitutional function of guiding the foreign affairs of government. The internal mechanisms of each government must be committed to giving the necessary priority to good-neighbour principles, since Latin America has already wasted decades in which declarations of principles held sway over bilateral and multilateral development plans.

The bilateral Amazonian cooperation plans have fulfilled the mandates proposed by the High Contracting Parties when they signed the Treaty for Amazonian Cooperation. The political will of governments has materialized in various bilateral agreements and treaties, which have unquestionably generated activities in the border areas distant from the centres of power.

The participation of the Organization of American States (OAS), as a continental and hemispheric forum, in guiding and preparing the bilateral development plans has been one of the most important, most practical elements of Amazonian policy under the Framework Treaty, since it awakened an interest on the part of member countries in regions that had traditionally been forgotten and abandoned to their sad fates.

In the case of Colombia, the implementation of the plans has been regarded as a priority objective of border policy. With the Ministry of Foreign Affairs acting as coordinator, the various government institutions involved in the development of the Amazon have focused sizeable budgetary and professional efforts on financing activities related to the development of the Amazon.

The greatest needs in the outlying areas are found in these parts of the country. Major bilateral advances have been achieved with the neighbouring countries of Ecuador, Peru, and Brazil, countries with which Colombia has bilateral Amazonian cooperation agreements.

The micro-integration of border areas was achieved with the important ingredient of the Amazonian culture, which institutionalized such significant concepts as the environmental policy described above.

The bilateral plans have created an Amazonian consciousness and have led many professionals to specialize in various related disciplines. These are the people in charge of developing new environmental rules, and in this work the OAS has offered continuous technical assistance.

The development of the overall Amazonian policy has found its greatest ally in the bilateral cooperation agreements, because it is along the Amazon borders, where the sovereignty of one nation ends and another begins, that the concept of a "frontier of cooperation" must be institutionalized and strengthened. This frontier has no limits, since the regions are inseparable in their geographical and anthropological features.

The experience and the important progress made under the bilateral Amazon cooperation plans in effect in the region have made an important contribution to the overall Amazon policy. It is the responsibility of the Treaty Secretariat Pro Tempore to coordinate them so that they become additional elements of planning that will help in obtaining outside funds to finance the implementation of the bilateral projects together with national counterpart funds. This important action will put new life into the Framework Treaty and make it possible for the agreed Amazonian policy to become a reality.

It is suggested that the regulations and efforts made by the member countries that have signed bilateral Amazonian cooperation plans be regarded as in the regional interest. Colombia made this proposal during its term as Secretariat Pro Tempore in 1989, and it believes that the time is ripe to allocate resources obtained through the action of the Treaty authorities to practical projects, such as those prepared with the technical cooperation of the Organization of American States.

3

The Amazon Cooperation Treaty: A mechanism for cooperation and sustainable development

Manuel Picasso Botto

Introduction

The main purpose of this document is to give an overview of the institutional scope and programmes of the Amazon Cooperation Treaty executed on 3 July 1978 in the city of Brasília by the Republics of Bolivia, Brazil, Colombia, Ecuador, Guyana, Peru, Surinam, and Venezuela.

For reference purposes, we will begin by explaining certain unique characteristics of the Amazon river and the Amazon basin, followed by the goals and objects of the treaty and the corresponding treaty bodies, closing with a summary of the major on-going programmes and projects within the frame of this international instrument.

Unique characteristics of the Amazon river and the Amazon basin[1]

– The Amazon region is known as the largest wet tropical forest. Its flora and fauna, including numerous plants and animals still unknown to science, make up more than 50 per cent of the global biota. It is estimated that over 56 per cent of the tropical forests of planet earth are found in the Amazon basin.

- The Amazon river, which runs through 6,762 km from its source in the snow-capped mountain of Mismi, in the department of Arequipa (southern Peru), is also the longest, largest, widest, and deepest river, and drains the largest basin in the world.
- The Amazon river discharges into the Atlantic ocean between 200,000 and 220,000 cubic metres of water per second, between 6.3 and 6.9 billion cubic metres per year, representing 15.47 per cent of all fresh water in the world. It also discharges approximately 1 billion metric tons of sediments per year.
- In the Obidos strait (Brazil), the Amazon reaches an approximate depth of 300 m, permitting access by large-draft vessels to Iquitos more than 2,300 km up river.
- The slope is very uneven in the upper basin, where it varies approximately 5,000 m along a straight course of 50 km. In the mid-lower stretch the gradient is less steep: from Iquitos (Peru) to its mouth the Amazon flows through 2,375 km with a gradient of only 4.5 cm per km.
- The width of the Amazon varies. During the rainy season, in some stretches the waters of the Amazon flood 20–50 km on both margins.
- The river bed extends over nearly 7,165,281 sq km. It accounts for 1.40 per cent of the total surface of the earth, 4.82 per cent of the continental surface, and 40.18 per cent of South America.
- The Amazon basin has very heterogeneous geographical and ecological characteristics, comprising territories located between 6,000 m.a.s.l. (the Andes mountains) and sea level. The lower basin is an immense biome of forests and waters. It is estimated that nearly 30 per cent of this region is formed by water and wet areas: rivers of diverse characteristics, lagoons, ponds, swamps, marshes, and flood-prone areas.
- The Amazon region provides undeniable global environmental services, such as control of the greenhouse effect, preservation of the hydric equilibrium of the atmosphere, circulation of nutrients, and conservation of biological, scientific, and cultural diversity.
- The Amazon is not an empty territory, although throughout most of the region the population density is low and nearly 60 per cent of it is urban. At present, in the lower areas there live approximately 22 million people assembled in native villages, modern forest populations engaged in extraction activities; and other populations, such as the Amazon population, grow at an annual rate of 3 per cent. There are nearly 379 ethnic groups, with a millenial

tradition of adaptation to the heterogeneous conditions of the region and depositories of an incalculable wealth of knowledge and technologies.

Objects and bodies of the treaty

The implementation of joint efforts and actions to promote the harmonious development of the respective Amazonian territories, in a way that permits environmental protection and the rational use of the natural resources from those territories, is one of the main objects of the treaty. To this end, the governments undertake to exchange information and to execute agreements and operational arrangements, as well as the relevant legal instruments.

Other objects of the treaty include the promotion of the harmonious development of the Amazon region, understood as the equilibrium between economic growth and preservation of the environment, improving at the same time the living conditions of the peoples of the region.

The treaty assigns special importance as well to actions aimed at the full incorporation of Amazonian territories to the national economies, rational utilization of water resources; improvement of navigable waterways, and the importance of establishing the adequate physical infrastructure between the member countries, especially in the areas of transportation and communications.

Improvement of the health conditions of the Amazonian population, as well as prevention and control of epidemics, close collaboration in scientific and technological areas, development of tourism without detriment to native cultures, sustainable use of natural resources, and conservation of regional, ethnological, and archeological resources are other goals that deserve special treatment. The scope of application of this sub-regional cooperation instrument is the Amazon basin and adjacent territories with very similar characteristics.

The treaty encompasses the following bodies:

(i)   The meeting of Ministers of Foreign Affairs is the highest body. It establishes the basic common policy guidelines, supervises and evaluates the general conduct of this cooperation process.

(ii) The Amazon Cooperation Council is formed by high-level diplomatic representatives of the contracting parties. It is responsible for supervising compliance with the goals and provisions of the treaty, and takes decisions on the performance of bilateral or multilateral studies and projects.

(iii) The Permanent National Commissions of each member country are responsible for the implementation in their respective territories of the provisions of the treaty, as well as of the decisions taken by the meetings of Ministers of Foreign Affairs and the Amazon Cooperation Council.

(iv) The treaty's Secretariat is assigned, on a rotary basis, to the Ministry of Foreign Affairs of one of the member countries to carry out the activities specified in the treaty and those ordered by the meetings of Ministers of Foreign Affairs and the Amazon Cooperation Council. It is also responsible for actions relating to international technical and financial cooperation.

(v) The Special Commissions of the Amazon region, devoted to the study of specific problems or subjects, are formed by entities designated by the member countries, responsible for the coordination, follow-up, and correct implementation in their respective countries of the approved programmes and projects and for the submission of new proposals of common interest to permit compliance with the actions specified in the treaty.

The Organizational Chart of the treaty is set forth in annex 1.

Meetings of Ministers of Foreign Affairs

Pursuant to Article XX of the treaty, the Ministers of Foreign Affairs of the contracting parties shall meet whenever deemed convenient or appropriate, to establish basic common policy guidelines, consider and evaluate the general course of the Amazon cooperation process and take decisions for the achievement of the goals proposed therein. These meetings are held at the initiative of any of the parties, supported by at least four other member states.

The general policies that guide the course of the treaty from the time that it was signed are therefore based on the directives issued by the five meetings of Ministers of Foreign Affairs of the member countries held thus far.[2]

The documents from the past five meetings of Ministers of Foreign Affairs of the treaty are: the Belem Declaration, signed at Belém do Pará, Brazil, on 24 October 1980; the Santiago de Cali Declaration signed at Cali, Colombia, on 8 December 1983; the San Francisco de Quito Declaration, signed at Quito, Ecuador, on 7 March 1989; the Santa Cruz de la Sierra Declaration, signed at Santa Cruz, Bolivia, on 8 November 1991, and the Lima Declaration signed at Lima, Peru, on 5 December 1995.

Upon the signing of the Lima Declaration, at the close of the Fifth Meeting of Foreign Affairs held at the initiative of the Government of Peru, it may be affirmed that the treaty entered a new promising consolidation phase: the decision to create a Permanent Secretariat and a new Special Commission of the Amazon Region on Education; the design for establishing a financial mechanism and an institutional network for research and protection of genetic resources, and the decision to negotiate an agreement on measures to prevent and control the contamination of water resources and its damaging effects have reshaped the institutional framework and enabled its prompt consolidation, lending even more vitality, cohesion, and continuity to this instrument, which has already acquired an internationally recognized profile and identity.

At this meeting, the first since the World Conference on the Environment and Development held at Rio de Janeiro in 1992, the ministers also discussed important topics and aspects relating to the Amazon forest, water resources, biological diversity, native peoples, environmental education, and hydrobiological resources, among other subjects.

The Amazon Cooperation Council

Pursuant to Article XXI, high-level diplomatic representatives from the Contracting Parties will meet annually as members of the Amazon Cooperation Council. The Council has the following responsibilities:
(i)   supervises compliance with the objects and goals of the treaty;
(ii) supervises compliance with the decisions taken by the meetings of Ministers of Foreign Affairs;
(iii) issues recommendations to the parties on the convenience or timeliness to hold meetings of Ministers of Foreign Affairs and prepares the respective agendas;
(iv) considers the initiatives and projects submitted by the parties and takes the relevant decisions for the performance of bilateral/multilateral studies and projects, to be implemented, as appropriate, by the Permanent National Commissions;
(v)   evaluates the performance of projects of bilateral/multilateral interest.
Formally, council members may hold ordinary and special meetings, both of which must be convened by the Pro Tempore Secretary. The delegations must be headed by a high-level diplomat from each

member country and composed of delegates, advisors, and other members accredited by the governments.

Until now, the Amazon Cooperation Council has met on seven occasions: in Lima, Peru, in July 1983; in La Paz, Bolivia, in September 1986; in Brasilia, Brazil, in March 1988; in Bogota, Colombia, in May 1990; in Quito, Ecuador, in July 1993; in Lima, Peru, in October 1994; and again in Lima, Peru, in November 1995. The eighth meeting took place in Caracas, Venezuela, during the first part of 1997.

Special Commissions of the Amazon region

Pursuant to Article XXIV of the treaty, seven Special Commissions of the Amazon region have been established for the study of the following specific problems and subjects: the environment; science and technology; transportation, communications and infrastructure; health; native affairs; tourism; and education. These Special Commissions are formed by the competent national institutions within each sector, linked into an active sub-regional communications network.

Special Commission on Science and Technology (CECTA)
CECTA was established during the Third Meeting of the Amazon Cooperation Council (Brasília, March 1988). Its purpose is to encourage and supervise the performance of regional projects and other activities undertaken by the member countries of ACT in the fields of science and technology; to act as a mechanism for obtaining funds from international sources and to coordinate the application of those funds in regional projects. CECTA has met on five occasions, the last two meetings were held in Lima in 1995 and Iquitos in 1996.

Special Commission on the Environment (CEMAA)
This Commission was established during the Third Meeting of Ministers of Foreign Affairs of ACT (Quito, March 1989), to carry through the common goals of environmental protection and rational use of the many and varied natural resources of the Amazon. The Commission has met on five occasions; its last two Annual Meetings were held in Lima in 1995 and in Santafe de Bogota in 1996.

Special Commission on Native Affairs (CEAIA)
CEAIA was established during the Third Meeting of Ministers of Foreign Affairs of ACT (March 1989), for the purpose of giving way

73

to the national interest in native affairs, in accordance with the following guidelines:

(i) to promote the strengthening of the ethnic identity and conservation of the historical and cultural heritage of native peoples, especially of their lands and resources;

(ii) to promote the exchange of information between the various bodies, institutes, and/or institutions responsible in each of the Amazonian countries for the design and implementation of national policies on native peoples, with the purpose of enhancing mutual knowledge on the native peoples of the region;

(iii) to foster technical cooperation programmes incorporated in the policies of member countries on native populations; and

(iv) to carry out programmes and projects of common interest in areas such as the conservation, management, and use of natural resources in native territories, the rescue and development of native technologies, community development, human resources training, etc.

This Special Commission has met four times; its last meeting was held in Lima in 1995.

Special Commission on Health (CESAM)

CESAM was created in March 1988 as a sectoral coordinating body within the Health area, during the Third Meeting of the Amazon Cooperation Council, with the following objects:

(i) encouraging, coordinating and supervising the implementation of regional programmes and other activities undertaken in the health area by the countries parties to the Amazon Treaty; and

(ii) acting as a mechanism to obtain international resources from international cooperation and coordinating the application of those resources in regional programmes.

CESAM has met on four occasions; its last meeting was held in Lima in June 1995.

Special Commission on Transportation, Infrastructure and Communications (CETICAM)

The Fourth Meeting of the Amazon Cooperation Council (Santafe de Bogota, May 1990) established the Special Commission of the Amazon Region on Transportation (CETRAM). Later, its mandate was expanded by the Fourth Meeting of Foreign Affairs of ACT held at Santa Cruz de la Sierra Bolivia, in November 1991, to cover communications and infrastructure, and its name was changed to Special Commission of the Amazon Region on Transportation, Infrastructure and Communications (CETICAM).

Generally, the programmes and projects of this Special Commission are designed to establish strategies for the enhancement of different modes of transportation; prepare general plans and projects on transportation along the Amazon river to develop commerce and contribute to regional prosperity; promote the establishment of a ground transportation system (by road and rail); and encourage regional air transport; facilitate telecommunications; carry out pre-feasibility and feasibility studies on major inter-oceanic corridors; and identify alternative intermodal connections between the basins of the Amazon, Orinoco, and Plata rivers.

The assessment of environmental impact of projects in the Amazon, particularly of projects relating to transportation infrastructure, has become at present a fundamental training component for the members of this Special Commission.

Special Commission on Tourism (CETURA)
CETURA was created during the Fourth Meeting of the Amazon Cooperation Council (Bogota, May 1990). Its main responsibilities are the following:
(i)   to promote training and organization of human resources and carry out market surveys on aspects related to tourism;
(ii) to promote the organization of meetings, fairs, and other activities to encourage ecological tourism in the Amazon;
(iii) to identify the effects of tourist-related activities upon natural resources and native communities, taking into account the particular environmental and cultural characteristics of the Amazon to incorporate them in the planning and development of projects;
(iv) to stimulate the complementary development of the special characteristics of each tourist site in the Amazon;
(v)   to promote the exchange of experiences, research, and studies on different areas of tourism in connection with investments, credit lines, etc. in the member countries; and
(vi) to encourage cooperation between national, regional, public, and private institutions engaged in tourism-related activities, in accordance with the general goals of the treaty.
This Commission has met on three occasions; it held its third annual meeting in Lima, in June 1995.

Special Commission on Education (CEEDA)

Two decisions concerning education were taken at the Fifth Meeting of Ministers of Foreign Affairs of the Amazon Cooperation Treaty: the Special Commission of the Amazon Region on Education (CEEDA) was created as a mechanism to evaluate experiences and coordinate the design of educational plans adapted to Amazonian circumstances, and to promote the training and organization of human resources in the Amazon region, as well as respect for the cultural identity of its peoples. It was also resolved to design a common programme to promote environmental education and awareness at school level. To this end this Special Commission was required to organize a regional seminar to propose the guidelines, contents, and scope of manuals for this purpose.

Through the creation of this Special Commission, the governments implemented a tool to permit the inhabitants of the Amazon basin to benefit from educational programmes and projects adapted to their needs, and to participate in forging their economic potential and environmental conservation.

The First Ordinary Annual Meeting of this Special Commission, as well as the Regional Seminar on the Preparation of Manuals on the Common Program on Promotion of Environmental Education and Awareness at School Level will take place in 1997, in accordance with the schedule of pending activities.

Pro Tempore Secretariat of ACT

The Pro Tempore Secretariat is responsible for carrying out the activities mandated by the Treaty, the Meetings of Ministers of Foreign Affairs and the Amazon Cooperation Council. It acts as a catalyst and promoter of efforts throughout the region, dynamizing the exchange of experiences, spreading scientific or technical information, and encouraging the formulation and implementation of regional projects. To this end, the Secretariat formulates and receives proposals; organizes and convenes seminars and workshops on specific topics; edits dissemination bulletins and publications; prepares projects to enhance knowledge about the region; promotes sustainable development in the Amazon; favours the conservation of biological diversity, and supports training for local populations.

The regulations of the Secretariat define its duties and responsibilities, namely:

(i)   to supervise compliance with the goals and objects of the Amazon Cooperation Treaty;

(ii)   to comply and supervise compliance with the resolutions adopted by the Meetings of Ministers of Foreign Affairs and the Amazon Cooperation Council;

(iii)   to coordinate with the competent authorities and bodies of the Amazonian countries details for the holding of meetings of formal and technical treaty bodies, to disseminate their results, and follow up the decisions taken thereby.

(iv)   to prepare, compile, and store the official correspondence of the Amazon Cooperation Treaty, which will be transferred at the time of making the corresponding rotation (duties are exercised on a rotary basis);

(v)   to keep the Permanent National Commissions of the treaty duly informed of the progress made by the various Special Commissions and the meetings and other activities of the Amazon cooperation process. The official communications between the Secretariat and the parties on the convening of technical meetings or meetings of the Special Commissions regarding substantive aspects of programmes, projects, and international cooperation are transmitted by diplomatic channels. The Pro Tempore Secretariat takes care that the Special Commissions observe the same procedure;

(vi)   it is charged, in accordance with the mandates of the Meetings of Ministers of Foreign Affairs, the Council and the Special Commissions, with the tasks of management, promotion, preparation of documents required for the financial negotiations, follow-up, and all that refers to the obtainment of resources for the financing of programmes and projects, as well as with the timely and efficient implementation and start-up of these programmes and projects. The projects, which may arise from initiatives of the Special Commissions or the Pro Tempore Secretariat, are submitted to all the parties for consideration by them, and their implementation is subject to prior and express approval of each of the concerned parties. The non concerned parties may present observations within a maximum term of 60 days.

(vii)   to apply for, arrange and submit proposals for technical, scientific, and financial cooperation for the approval of the member countries;

(viii) to follow-up and make a general evaluation of on-going programmes and projects, and take appropriate measures for their
timely and efficient implementation;

(ix) to protect and update all the documents concerning progress
made in implementation of the treaty, bilateral agreements on
Amazonian cooperation, and related instruments;

(x) to disseminate information continuously on the cooperation
process, for the purpose of attracting positive attention from
international organizations, other countries, and public and
private organizations;

(xi) to prepare, in coordination with the Ad Hoc Consultative
Committee and regional coordinating bodies of the Special
Commissions, an annual workplan including the proposals for
operational research plans and implementation of specific programmes and projects. The Secretariat must submit for approval annually the Work Schedule of the Special Commissions to
the parties, and strive to hold meetings of all the Commissions
prior to the Meeting of the Council in order that the latter may
perform an adequate follow-up of their activities;

(xii) to submit detailed reports biannually and at the end of its
period of office;

(xiii) to coordinate the activities of the Special Commissions of the
Amazon Region with the respective Executive Secretariats;

(xiv) to convene, at the request of any of the member countries, the
necessary technical meetings for coordinating the actions of the
various bilateral and multilateral mechanisms of the treaty, and
to deliver to the Amazon Cooperation Council reports on the
results achieved, for the purpose of facilitating the evaluation of
the Amazon cooperation process;

(xv) other duties and responsibilities that may be assigned to the
Secretariat by the Meetings of Ministers of Foreign Affairs and
the Amazon Cooperation Council.

The duties of the Pro Tempore Secretariat are exercised on a rotary
basis by the Ministries of Foreign Affairs of the Countries Parties to
the Amazon Cooperation Treaty. The respective Ministry of Foreign
Affairs designates a high-level diplomatic officer to act as Pro Tempore Secretariat, supported by a team of diplomatic officers devoted
full-time to this task.

Since the treaty was signed, the duties of the Secretariat have
been exercised by the following countries: Peru, from October 1980
to July 1983; Bolivia, from July 1983 to September 1986; Brazil, from

September 1986 to March 1988; Colombia, from March 1988 to May 1990; Ecuador, from May 1990 to January 1994; and Peru, since February 1994. This responsibility will shortly be transferred to the Ministry of Foreign Affairs of Venezuela.

The Pro Tempore Secretariat is assisted by the Ad Hoc Consultative Committee, composed of the Ambassadors of the governments of the parties accredited before the country in charge of the Secretariat. It is the duty of the Secretariat to exchange information and coordinate the actions relating to the Amazon cooperation process with this committee.

Specialists in each specific area of concern of the Special Commissions work at the Secretariat's headquarters and act as regional coordinators. It is their responsibility to promote regional projects; to implement actions that facilitate the process of exchange of information, to prepare technical documents, to update regional data banks, and to establish the necessary contacts for wider participation of institutions and experts at seminars, workshops, and meetings.

In recent years, the Secretariat has promoted a vigorous institutional mobilization of entities and sectors responsible for the policies and actions of the member countries in the Amazon region, reflected in a substantial growth of participation in the cooperation process that has benefited the treaty. The Secretariat has acted as a catalyst in coordinating the adoption of a common stand by the countries parties to the ACT at international fora devoted to the discussion of matters of interest to the parties. Likewise, the Secretariat has formulated several projects of regional scope involving a total investment of more than US$25m, with the support of prominent bodies of the United Nations, for some of which the financing arrangements are at an advanced stage.

The organizational chart of the Secretariat during the tenure of office of Peru is attached as annex 2.

Workplan

The Secretariat's workplan, approved by the Foreign Ministers at their Fifth Meeting, contains programmes, projects, and activities oriented at and designed for achieving the goals of the treaty, especially with reference to improving the quality of life of the population.

In accordance with the policy guidelines issued by the Meetings of Ministers of Foreign Affairs and the Amazon Cooperation Council, the workplan points at the adoption and application by the member countries of policies and strategies for sustainable development based on previously identified and common needs, prospects, and priorities.

(a) In the political and institutional areas, the workplan aims, among other things, to achieve the following objectives:

(i) to contribute to the reinforcement of the treaty's institutional base in each member country, supporting the action of the Permanent National Commissions;

(ii) to adopt clear mechanisms for the management of the Secretariat, so as to permit a fluid exchange and communication with the parties and to ensure due performance of its duties;

(iii) to strengthen the work of the Special Commissions of the Amazon region;

(iv) to foster the exchange of views between the member countries for the adoption of common positions at various international fora on matters relating to the treaty, and to achieve closer relations with different sub-regional bodies and mechanisms for the treatment of matters of common interest;

(v) to promote the adoption of agreements on specific subjects for the preservation of the environment;

(vi) to promote coordinating action between the political-diplomatic bodies of the treaty through the Special Commissions of the Amazon Region in the process of formulation of projects.

(b) In the technical area:

(i) to strive for the joint preparation of projects with a regional focus;

(ii) to lay down criteria for the identification and selection of projects, as well as follow-up mechanisms;

(iii) to evaluate the potential and limitations of previously identified projects, and to determine their priority level in terms of the common needs of the member countries.

This workplan is translated into specific items in the Operational Plans of each of the Special Commissions of the Amazon Region, which generally respond to the following objectives:

- In the areas of science and technology, to contribute to the development of regional policies and strategies based on those laid down by the countries parties to ACT, for the consolidation of science and indigenous technologies conducive to viable alternatives for the sustainable development of the Amazon region.

- Within the scope of action of the Special Commission on the Environment, to develop and implement environmental protection strategies, through the adequate management of natural resources, to permit an ecological equilibrium that guarantees the sustainable development of the Amazon region; and to encourage research on and dissemination of the efficient use of Amazonian biodiversity for the sustainable development of the region, promoting the exchange of experiences on the natural productivity of ecosystems, species, and genetic resources;

- on the subject of native affairs, to promote the adequate treatment of the problems of native populations of the Amazon, striving to safeguard and re-evaluate their ethnical and cultural heritage and ensure the protection of their lands;
- in the areas of transportation, communications and infrastructure, to have the member countries adopt and apply coordinated policies and strategies environmentally compatible with the goals of sustainable development of the Amazon region; and to incorporate the Amazonian territories to the economic activities of their respective countries, as well as to the integration efforts promoted through bilateral and multilateral actions in the context of the Amazon Cooperation Treaty;
- in the area of tourism, to strive towards the establishment of a common policy in the member countries for attracting tourism to the Amazon region without affecting the environment and the native populations, and to incorporate those populations insofar as possible into the activities to be implemented;
- in the area of health, to contribute to the improvement of the health conditions of the populations of the Amazon, with special emphasis on native communities, by strengthening primary health care programmes to provide adequate health services to Amazon populations, attending to the characteristics of the Amazon tropical climate and to the cultural heterogeneity of the Amazonian population, in terms of specialized health and nutritional services.

The programmes, projects, and technical activities are carried out through the Special Commissions and Permanent National Commissions. The Pro Tempore Secretariat is in charge of the follow-up and general evaluation of on-going programmes and projects and takes the necessary actions for their prompt and efficient execution.

Eight programmes have been identified within the scope of the Special Commissions, and each member country has been assigned regional cooperation responsibilities. In the specific case of the programmes of CEAIA, there are only seven; the programmes of CETURA are coordinated region-wide by the Secretariat. Those programmes are classified according to topics and comprise related projects and activities. A table detailing the various programmes is included as Annex 3 below.

With regard to the scope of the projects, the aim is to encourage the use of human resources and infrastructure existing in each country, and to implement projects of wide scope and high cost in stages to facilitate their financing. Another aim is to combine activities with training and educational components, dissemination of information, experiences, research, and studies.

Many of these activities are developed with funds from non reimbursable international cooperation. The member countries have yet to make regular contributions of their assigned regular quotas, although they do contribute human and financial resources for the implementation of national activities; each government acting as seat of the Pro Tempore Secretariat assigns funds for the installation and operation of the Secretariat.

The international cooperation provided by organizations and governments allows the treaty to carry out projects, especially in relation with the identification of potentials and natural resources management. It also supports the activities of the Pro Tempore Secretariat and the Special Commissions, permitting the exchange of experiences and holding of meetings, and translation in many cases of various national initiatives into regional projects. This process has activated and stimulated the work of the government institutions and sectors that make up the Permanent National Commissions of ACT and facilitates smooth and mutual inter-institutional relations between the parties.

Various international organizations have been providing support for the treaty in the form of non reimbursable multilateral assistance, namely: the European Union, the United Nations Development Programme (UNDP), with funds from the Global Environment Facility (GEF), the United Nations Food and Agricultural Organization (FAO), with funds from the Kingdom of the Netherlands; the Inter-American Development Bank, the World Bank, the United Nations Industrial Development Organization (UNIDO); the Organization of American States (OAS), the Corporacion Andina de Fomento (CAF) (Andean Development Corporation), among others. The Governments of Finland and Canada have recently joined this flow of cooperation.

The Pro Tempore Secretariat has established similar links with prominent institutions through the signing of memoranda of understanding with the United Nations Environmental Programme, the

Asociacion de Universidades Amazonicas (UNAMAZ) (Association of Amazonian Universities), the World Resources Institute (WRI), among others.

Principal ongoing programmes and projects

Within the frame of CEMAA
The Regional Strategies Project for Sustainable Conservation and Management of Natural Resources in the Amazon comprises three projects: "Zonificacion Ecologica y Monitoreo Geografico en el Amazonas" (ecologic zoning and geographic monitoring of the Amazon); "Capacitacion en el Aprovechamiento Sustentable de la Biodiversidad Amazonica" (training in sustainable uses of Amazonian biodiversity); and "Manejo de Recursos Naturales en Territorios Indigenas de la Amazonia" (natural resources management in native Amazonian territories). Their implementation is financed by UNDP-GEF, to contribute to the ordering of the territory as well as to the study, formulation of strategies, and evaluation of Amazonian biodiversity.

The Program on Ecologic Zoning and Geographic Monitoring of the Amazon, funded by IDB, covers two sub-programs: Technical Assistance for the Permanent National Commissions of the Amazon Cooperation Council and Assistance for the Harmonization of Geographical Information Systems (GIS). This last sub-programme complements the zoning project, which began with funds furnished by UNDP, aimed at harmonizing the methodological proposals of the aforementioned projects.

The projects on ecologic zoning and geographical monitoring of the Amazon aim to have the countries parties to ACT adopt and apply ecologic-economic zoning principles in the formulation of policies and strategies for ordering their territories, as basic tools and decision-making criteria, to contribute thus to the sustainable development of the Amazon region, the strengthening of national zoning institutions, and the identification of alternatives for harmonizing the hardware and software installed in the ACT member countries to make their national geographical information systems mutually complementary.

The Regional Project on Planning and Management of Protected Areas in the Amazon Region is funded with resources from the European Union and its purpose is to ensure the conservation of natural and cultural biodiversity by consolidating selected pilot areas and demonstration centres and establishing protected areas in the Amazon. This project also oversees, among other activities, the design of management plans for ecotourism. It is important to highlight, in this regard, that given the successful results obtained thus far in implementation, the parties have agreed to carry out the second stage of the project.

The following projects and research activities are being carried out with funds provided by FAO: Food Security, Nutrition and Natural Resources of the Amazon; Biodiversity and Food in the Amazon; Recovery of Native Foods; Nutritional Alternatives; Propagation of Promising Plant and Vegetable Species in the Amazon Region; Small Agrobusiness as a Factor of Sustainable Development in the Amazon Region; Sustainable Use and Conservation of Wild Fauna in the Countries of the Amazon Basin; Pilot Project on Management of Podoenemis Expansa; Forests, Trees and Rural Communities; Identification of Criteria and Indicators of Sustainability for the Amazon Forest, etc. These projects refer mainly to the knowledge, management, and evaluation of natural resources.

Finally, technical assistance from UNIDO is being used to carry out the project "Strategy for Environmental Quality Management in Three Member States of ACT: Bolivia, Ecuador and Peru," aimed at laying down policies for incorporating environmental and industrial factors into developmental actions undertaken in the Amazon region.

Within the frame of CETICAM
In accordance with the treaty provisions on the convenience of creating an adequate physical infrastructure between the member countries, especially in the areas of transportation and communications, member countries are developing the project "Transportation Network for the Amazon Region," as a basic planning tool and physical space for the implementation of regional transportation policies, environmentally compatible with the goals of sustainable development for the Amazon region, and as a general framework for the development of various projects stemming from the integration needs of the region.

The first stage of this project foresees the formulation of a proposal on an Intermodal International Transportation Network for the Amazon Region, supportive of and environmentally compatible with the goals on sustainable development advanced for the region, and contributing to:
(i)   the integration efforts of the member parties, promoted through

actions performed within the framework of the Amazon Cooperation Treaty;

(ii) the full incorporation of their Amazonian territories into the scope of activities of their respective national economies; and

(iii) generation of a physical infrastructure compatible with the aspirations of the inhabitants of the Amazon region and the access needs of isolated population centres.

This first stage of this project has been funded by the European Union and implemented by the Empresa Brasileña de Planeamiento de Transportes (GEIPOT) (Brazilian Transportation Planning Enterprise).

The studies for the formulation of this proposal have been completed and the working draft was approved by the representatives of the member countries during a technical workshop organized by the Pro Tempore Secretariat and held in Caracas 3–6 December 1996.

The final version of the network, including specific information provided by the governments of the parties, is being prepared and will be circulated. The Fourth Ordinary Meeting of CETICAM held at Caracas approved this proposal.

Within the frame of CECTA

Funds provided by the World Bank's Economic Development Institute (EDI), the Government of the Netherlands, FAO, and UNDP are being used to carry out the project "Dissemination of Sustainable Technologies for the Use of Amazonian Biodiversity," and to publish important and widely disseminated technical documents within the frame of this project, such as: Diagnostico de los recursos hidrobiologicos de la Amazonia (diagnosis of hydrobiological resources of the Amazon); Experiencias agroforestales exitosas en la Amazonia (succesful agroforest experiences in the Amazon); Recursos fitogenéticos de cultivos alimenticios y frutales amazónicos (phytogenetic resources of Amazonian food crops and fruits); Plantas medicinales de la Amazonia: realidad y perspectivas (Amazonian medicinal plants: reality and prospects); Biodiversidad y salud en las poblaciones indigenas de la Amazonia (biodiversity and health among Amazonian native populations); Uso y conservacion de la fauna silvestre en la Amazonia (use and conservation of Amazonian wild fauna); Patentes, propiedad intelectual y biodiversidad amazonica (patents, intellectual property and Amazonian biodiversity); and Cultivo del pijuayo para palmito en la Amazonia (production of "pijuayo" for palmetto in the Amazon).

The document "Frutas y hortalizas promisorias de la Amazonia" (promising Amazonian fruits and vegetables) was recently published with technical and financial support from FAO. This work describes 52 plant species, and constitutes a valuable contribution for the study and knowledge of a series of plant products with economic and social potential by public and private institutions, researchers, technicians, and persons interested in the Amazon region and its incorporation to the local, national, and regional economies.

Within the frame of CEAIA
The project entitled "Programa Regional de Consolidacion de Territorios Indigenas a traves del Tratado de Cooperacion Amazonica" (Regional Program for the Consolidation of Native Territories through the Amazon Cooperation Treaty) was implemented with technical and financial support from the European Union, with the general aim of assisting certain Amazonian native communities of Bolivia, Ecuador, and Peru in the process of surveyance and legalization of the lands inhabited by them, and laying down conditions that may permit those communities to manage, preserve, and make rational use of the natural resources stored therein.

A regional diagnosis was prepared under this project and guidelines were issued on native lands, embodied in the recent publication of the Secretariat entitled "Tierras y aguas indigenas de la Amazonia: una experiencia regional" (native lands and waters of the Amazon: a regional experience).

The parties to the treaty commissioned the Secretariat, at the Fourth Ordinary Meeting of CEAIA held in Lima in May 1995, to formulate a new regional project for promoting regional support for processes of recognition and protection of native lands and waters in the Amazon, and for the sustainable management of biological resources. In this context, the Secretariat, in coordination with the national entities responsible for native affairs in each of the member countries, identified the respective national proposals and prepared a basic proposal for a regional project entitled "Programa Regional de manejo sostenible de recursos naturales en areas indigenas de la Amazonia" (Regional programme for sustainable management of natural resources in native areas of the Amazon), which was submitted to the parties for consideration.

Bearing in mind that efforts for the systematization of information on native Amazonian populations within the frame of the Amazon Cooperation Treaty are still incipient, the Pro Tempore Secretariat through the Coordination of the Special Commission of the Amazon Region on Native Affairs (CEAIA) proposed at the Fourth Ordinary Meeting of CEAIA to include the subject of native affairs within the system of SIAMAZ and not to establish a separate information subsystem. Based on this perspective, scattered public information is being collected and will shortly be presented in CD-ROM format.

Within the frame of CESAM
The process of formulation and consultation of the regional project "Promocion de la Salud en las Poblaciones de la Region Amazonica" (Health promotion among Amazonian populations) is being completed, involving financial aid from the European Union for the amount of US$1,821,396 and contributions from the concerned countries equivalent to US$1,278,089. The main object of this project is to reinforce community health services in the Amazon region through the development of a regional programme of training, technical assistance, and exchange of national experiences in health promotion in the Amazon.

The specific goals of this project should lead to the establishment of a pilot regional health programme designed to strengthen and foster cooperation in the area of health, and to raise the levels of capacity and management of health personnel, among other objectives.

Within the frame of CETURA
During the Fifth Meeting of Ministers of Foreign Affairs, the Secretariat was commissioned to formulate, in consultation with the parties, the Master Development Plan on Tourism and Ecotourism for the Amazon Region, as an instrument for promoting regional development and investments in tourism in the Amazon, with the support and active participation of public and private sectors engaged in tourism, environmental, and native affairs of the member countries.

The plan is being formulated on the basis of the regional project "Promicion del Desarrollo del Ecoturismo en la Region Amazonica" (Development of ecotourism in the Amazon region), designed during the Third Ordinary Meeting of CETURA held in Lima in June 1995, and the regional workshop that preceded it. The objective of this new proposal is to cover not only ecotourism, but also sustainable tourism as an activity that subsumes ecotourism.

Both the master plan and the regional project will be examined by the authorized representatives of the member countries at the Fourth Meeting of CETURA to be held in the city of Manaos.

Within the frame of CEEDA

The activities to be carried out by this Special Commission are the holding of its First Ordinary Meeting and the preparation, in accordance with the commission received from the Fifth Meeting of Foreign Affairs Ministers, of a common programme for the promotion of environmental education and awareness at school level, and, to this end, the organization and convening of a regional seminar to propose the main lines, contents, and scope of manuals.

Scheduled activities

The workplan of the Pro Tempore Secretariat includes, among other things, development of the following activities:

(i)   to design plans and strategies for soil conservation and improvement in the region, adequate soil use and management, and promotion of new productive activities based on native species of flora, fauna, and micro-organisms;

(ii) to implement a programme for environmental education and awareness at school level, and to begin with book production for this purpose;

(iii) to negotiate a framework agreement on actions to prevent and control the contamination of shared water resources and its negative impact on human health, the habitat, and biological diversity as a whole;

(iv) the future negotiations of the Amazonian countries shall take into consideration the unique characteristics of the region. For this purpose, it may be necessary to evaluate the methods applied in other river basins, as in the basins of the Mekong, Senegal, Plata, and Rhine rivers. It will also be important to have available updated studies on the problem of contamination levels in the Amazon basin;

(v)   to analyse and adopt systems for the registration of Amazonian genetic resources, and rules on protection of intellectual property and similar rules on protection of traditional knowledge, as well as on access to and intellectual property of Amazonian biogenetic resources;

(vi) to implement an institutional network for the protection of and research on genetic resources;

(vii) to follow a systematic approach to lay down common rules in the member countries on the sustainable use of the Amazon forest, and, in this connection, to support the initiatives of the member countries to implement the so-called "Tarapoto process" on sustainability criteria and indicators for the Amazon forest. To this end, the Secretariat has been able to obtain funds with international cooperation and to sponsor the holding of national evaluation workshops, one already held in Colombia, and others held in Peru and Ecuador. Likewise, the following actions are underway to expand regional cooperation:

(a) To conclude the reformulation and negotiation of financing by GEF of the project provisionally entitled "Accion para una Amazonia sostenible" (Action for Amazonian sustainability), and begin its implementation.

(b) To complete the process of formulation and consultation of the following projects: "Uso sostenible y conservacion de la fauna silvestre en los paises de la cuenca del Amazonas" (Sustainable use and conservation of wild fauna in the countries of the Amazon basin); "Programa regional de desarrollo y promocion del turismo sostenible y ecoturismo en la region Amazonica por intermedio de las comunidades locales" (Regional programme for development and promotion of sustainable tourism and ecotourism in the Amazon region through local communities); and begin the design of the second stage of the regional project "Planificacion y manejo de areas protegidas de la region Amazonica" (Planning and management of protected areas in the Amazon region).

(c) To continue the work of defining the scope of transportation programmes under the Special Commission of the Amazon Region on Transportation, Communications and Infrastructure (CETICAM) and of the following projects: "Estrategias para promover corredores prioritarios de transporte en la Amazonia y estrategias para su ejecucion por partes" (Strategies to promote priority transportation corridors in the Amazon and phased implementation strategies); "Red de telecomunicaciones para la region Amazonica" (Telecommunications network for the Amazon region); and "La cuenca Amazonica y su factibilidad de interconexion con las cuencas de los Rios Orinoco y de La Plata" (The Amazon basin and feasibility of its interconnection with the Orinoco and Plata river basins).

(d) To pursue the task of defining the scope of and formulating proposals for the following projects identified by the meetings of the Special Commissions, workshops, and other technical meetings: "Consolidacion, manejo y aprovechamiento de recursos naturales en areas indigenas de la Amazonia" (Consolidation, management and use of natural resources in native areas in the Amazon); "Sistemas integrales de produccion para el desarrollo agrario de la Amazonia" (Integral production systems for agricultural development in the Amazon); "Levantamiento de palmeras de la region Amazonica con potencial economico y social" (Survey of palm trees of the Amazon region with economic and social potential); "Experiencia piloto en el manejo de la fauna silvestre en los paises de la cuenca Amazonica" (Pilot experience in the management of wild fauna in the countries of the Amazon basin); "Plan regional de manejo para la conservacion y uso sostenible del caiman negro" (Regional management plan for the conservation and sustainable use of black cayman); "Recuperacion de la productividad de los suelos afectados por la agricultura y ganaderia mediante el uso de terrazas y sistemas agroforestales y silvopastoriles intensivos" (Recovering the productivity of soils damaged by agriculture and livestock production through the use of terracing and intensive agroforest and jungle grazing systems); "Plan referencial de homogenizacion de los sistemas de control epidemiologico para los paises del Tratado de Cooperacion Amazonica" (Reference plan for standardization of epidemiological control systems in the countries of the Amazon Cooperation Treaty); "Prevencion de daños producidos por la contaminacion con mercuriales" (Prevention of damage caused by mercury contamination).

Dissemination

For its dissemination activities on topics relating to development indexes in the Amazon region, the wealth of biological diversity and natural resources, the inventory of projects under the treaty, legislation in force, bibliography, publications, studies, and other data on the environment, health, science and technology, transportation, communications, native affairs, tourism, and education in the Amazon, as well as research centres in and for the region, the Secretariat has used the following means to carry out this task efficiently.

Preparation of trilingual quarterly information bulletins and their distribution to government entities of the member countries, academic and research institutions interested in the Amazon, regional or state governments, members of National Congresses or Parliaments, cooperating governments and agencies, NGOs, the media, private enterprises interested in the development of the Amazon, participants at workshops and seminars organized or sponsored by the Pro Tempore Secretariat of the Amazon Cooperation Treaty.

Likewise, a Website has been placed at the service of interested governments, researchers, and organizations within the Internet system. Through this service, the Secretariat furnishes the information contained in the 10 information bulletins published thus far. There are plans to expand this service, which will soon present its most important publications. It has also begun the task of compiling all the publications issued to date by the Secretariat in CD-ROM format.

Finally, a significant number of publications has been issued by the Secretariat – the complete list is set forth in annex 4 – dealing with subjects of interest to the treaty. There is a prior mechanism of coordination and consultation for all publications in which all member countries participate, allowing adequate time for the parties to comment on the contents of the proposed publications.

Conclusions

It may be affirmed today that the treaty has a joint work programme in specific areas that involves the participation of increasingly more and diverse institutions from each of the eight member countries.

The Secretariat has strived to define paths, to reinforce consultations, to expand institutional participation and to present the treaty as an effective instrument for channelling regional initiatives into research in the sustainable development of the Amazon basin. To a certain extent, some of these goals have been attained.

The periodic holding of meetings of the treaty mechanisms has facilitated the adoption of decisions that support, reinforce, and expand the scope of common action. More than 40 meetings, seminars, or workshops have been held by the parties recently, to discuss substantial policy matters as well as specific technical topics. All of them have or will have concrete effects on the sustainable economic yield of Amazonian resources, improvement of the living standards of the population, reinforcement of the institutional network or on the political dialogue at governmental level.

The government signatories of the treaty, in accordance with decisions taken at the Fifth Meeting of Foreign Affairs Ministers, will shortly install a Permanent Executive Secretariat of the treaty in Brasília. An ad hoc working group was established to that end, and convened on various occasions to meet both in Brasília and in Lima to prepare a proposal that will be duly considered by the competent bodies. The change of status of the Secretariat entails a modification in the text of the treaty, which will be considered in due course by the respective National Congresses.

There is still a long road ahead for achieving full compliance with the treaty goals. The political will demonstrated by the parties shows that it is fully possible to reconcile interests and opinions for carrying into effect development plans compatible with sustainability criteria advanced in 1978 and internationally sanctioned by the Rio Summit on Sustainable Development in 1992.

There are still, however, huge and important challenges ahead. The future of the Amazon region, a subject on which there persist certain differences, must be gradually approached while the experience of horizontal cooperation continues to be consolidated. The results of this process, based on dialogue and exchange of information, should have a favourable impact on the adoption of common policies regarding technical cooperation, the use and conservation of natural resources and implementation of projects. Gradually, the result should be universal agreement on criteria and concepts, in order to arrive at a system of physical data acceptable to all for understanding the complex character of biological diversity as a transborder phenomenon.

This process will be strengthened when, as a result of the evaluation of the projects and actions that are implemented, the accumulated experiences are systematized; common operational criteria for the projects are defined; sustainable development models adapted to the unique characteristics of the Amazon are applied; and sustainable development policies and their national, regional, and international inter-relations are analysed and updated.

I thank this Forum, jointly organized by the Consejo Mundial de Aguas, the Asociacion Internacional de Recursos Hidricos y la Organización de Estados Americanos (Organization of American States (OAS)) for the opportunity of explaining the mechanisms and bodies of the Amazon Cooperation Treaty, as well as for the level of operativeness and coordination it demonstrates today. I trust that the exchange of information on the basins of the Plata, São Francisco, and Amazon rivers will contribute important concepts for all Latin America, and specifically for the representatives of the member countries grouped in CETICAM, which has, among others, the task of identifying alternatives of intermodal connection between the basins of the Amazon, Plata, and Orinoco rivers, taking into account the waterways and the physical characteristics of the region.

This task implies giving priority to the utilization of the extensive network of waterways in these basins and maximum intermodal integration with operative roads, to prevent the risk of negative environmental impact caused by new roads, with the consequent deforestation of huge areas and occupation of unsuitable land for agricultural purposes.

The objectives of this programme are consistent with other programmes of the Special Commission, such as the one on river transportation, aimed at designing general strategies and specific projects for the development of river transportation as a natural means of access and communication in the Amazon basin, where the Amazon river is the backbone of transportation.

Notes

1. D'Achile, Barbara, Brack Egg, Antonio, and Wust, Walter H., 1996. *Uturunkusuyo: El Territorio del Jaguar. Perú: Parques Nacionales y otras áreas de conservación ecológica*, 1st edn. Lima: De Peisa, Banco Latino.
2. In addition, the Presidents of the Amazonian countries have met twice in the city of Manaos, upon the initiative of the Government of Brazil: in May 1989 and in February 1992, at which times the Presidents issued the "Declaration of the Amazon" and the "Documento de Posicion Conjunta de los Paises Amazonicos con miras a la Conferencia de las Naciones Unidas sobre el Medio Ambiente y el Desarrollo" (Joint Document of the Amazonian Countries towards the United Nations Conference on Environment and Development).

Annex 1   **Organizational chart of the Amazon Cooperation Treaty**

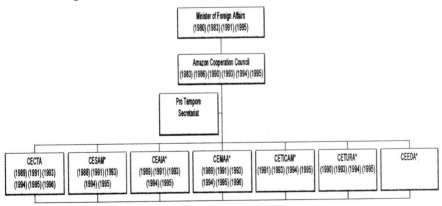

*Permanent National Commissions of the ACT

Annex 2   **Pro Tempore Secretariat of the ACT (1994–1997)**

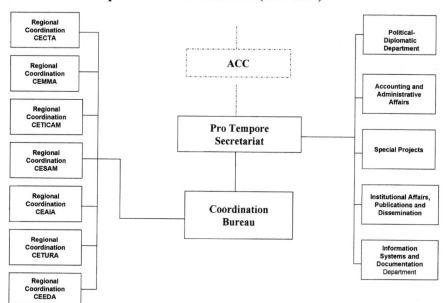

Annex 3   **Regional programmes within the framework of the Amazon Cooperation Treaty**

I. Special Commission of the Amazon Region on the Environment

CEMAA-1 "Zonificación ecológica-económica y monitoreo de las alteraciones en el uso de la tierra"
Coordinator: Brazil

CEMAA-2 "Ecología, biodiversidad y dinámica de poblaciones"
Coordinator: Venezuela

CEMAA-3 "Fauna silvestre"
Coordinator: Suriname

CEMAA-4 "Recursos hidrobiológicos"
Coordinator: Perú

CEMAA-5 "Defensa y aprovechamiento de los recursos forestales"
Coordinator: Ecuador

CEMAA-6 "Planificación y manejo de areas protegidas"
Coordinator: Colombia

CEMAA-7 "Unificación y/o interrelación metodologías para la evaluación de impactos ambientales, compatibilización de legislaciones ambientales e intercambio de informaciones sobre programas nacionales de protección del medio ambiente en la región amazónica.
Coordinator: Bolivia

CEMAA-8 "Investigación animal"
Coordinator: Guyana

II. Special Commission of the Amazon Region on Science and Technology

CECTA-1 "Inventario, uso, manejo y conservación de suelos"
Coordinator: Guyana
CECTA-2 "Sistemas integrales de producción vegetal"
Coordinator: Ecuador
CECTA-3 "Balance hídrico"
Coordinator: Bolivia
CECTA-4 "Estudios e investigaciones en ciencias sociales"
Coordinator: Colombia
CECTA-5 "Planificación y gestión en ciencia y tecnología"
Coordinator: Brazil
CECTA-6 "Producción animal"
Coordinator: Suriname
CECTA-7 "Desarrollo y adaptación de tecnologías para la amazonia"
Coordinator: Venezuela
CECTA-8 "Recursos zoo y fitogenéticos"
Coordinator: Perú

III. Special Commission of the Amazon Region on Health

CESAM-1 "Planificación general de la salud y sistema regional de información"
Coordinator: Bolivia
CESAM-2 "Materno-Infantil y atención primaria"
Coordinator: Brazil
CESAM-3 "Enfermedades tropicales"
Coordinator: Ecuador
CESAM-4 "Saneamiento básico"
Coordinator: Perú
CESAM-5 "Desarrollo y organización de servicios de salud"
Coordinator: Venezuela
CESAM-6 "Desastres"
Coordinator: Colombia
CESAM-7 "Medicina tradicional y salud en las comunidades indígenas"
Coordinator: Ecuador
CESAM-8 "Medicamentos básicos, escenciales y genéricos"
Coordinator: Guyana

IV. Special Commission of the Amazon Region on Indigenous Affairs

CEAIA-1 Conocimiento de culturas indígenas
CEAIA-2 Participación indígena en programas que le afecten
CEAIA-3 Atención Estatal a las comunidades indígenas
CEAIA-4 Educación en las comunidades indígenas
CEAIA-5 Salud en las comunidades indígenas
CEAIA-6 Legislación indígena

CEAIA-7 Desarrollo Regional y las Comunidades Indígenas
CEAIA-8 Coordinación de programación y sistema regional de información

V. Special Commission of the Amazon Region on Transportation, Infrastructure, and Communications

CETICAM-1 "Plan general de transporte para la región amazónica"
Coordinator: Brazil
CETICAM-2 "Transporte fluvial"
Coordinator: Bolivia
CETICAM-3 "Transporte terrestre"
Coordinator: Colombia
CETICAM-4 "Transporte aéreo"
Coordinator: Perú
CETICAM-5 "Comunicaciones"
Coordinator: To be decided
CETICAM-6 "Corredores interoceánicos"
Coordinator: Ecuador
CETICAM-7 "La Cuenca Amazónica y la factibilidad de interconexión con las cuencas de los ríos La Plata y Orinoco"
Coordinator: Venezuela
CETICAM-8 "Infraestructura"
Coordinator: To be decided

VI. Special Commission of the Amazon Region on Tourism

CETURA-1 "Plan de desarrollo turístico integrado y sistema de información turística de la subregión amazónica"
Coordinator: Suriname/Guyana
CETURA-2 "Investigaciones de mercado, comercialización y promoción turística conjunta de la subregión amazónica"
Coordinator: Venezuela
CETURA-3 "Apoyo a la información y capacitación turística de los recursos humanos de la subregión amazónica"
Coordinator: Perú
CETURA-4 "Compatibilización de la legislación turística y facilitación turística de los para la subregión amazónica"
Coordinator: Colombia
CETURA-5 "Evaluación de los efectos del turismo en el desarrollo sostenible de la subregión amazónica"
Coordinator: Bolivia
CETURA-6 "Identificación de oportunidades de inversión, promoción de inversiones en servicios de infraestructura turística con contenido de integración"
Coordinator: Brazil
CETURA-7 "Investigación en materia de infraestructura turística en la subregión amazónica"
Coordinator: Ecuador

Annex 4   List of Publications edited by the Pro Tempore Secretariat of the Amazon Cooperation Treaty (1994–1997)

Issued

Diagnóstico de los recursos hidrobiológicos de la Amazonia (Diagnosis of the hidrobiological resources in the Amazon region). Amazon Cooperation Treaty, Pro Tempore Secretariat. Lima, Perú. 162 pp. 1994. (SPT-TCA/No. 22)

Experiencias agroforestales exitosas en la Cuenca Amazónica. (Successful agroforestal experiencies in the Amazon basin) Amazon Cooperation Treaty, Pro Tempore Secretariat. Lima, Perú. 195 pp. 1994. (SPT-TCA/No. 23)

Propostas de políticas e estratégias regionais para o aproveitamento sustentável dos recursos fitogenéticos de cultivos alimentíceos e fruteiras amazônicas. (Political proposals and regional strategies for the management of the fitogeneric resources of amazon nutritional crops and fruits) Amazon Cooperation Treaty, Pro Tempore Secretariat. Lima, Perú. 49 pp. 1994. (SPT-TCA/No. 24)

Manual de entrenamiento: Sistemas fotovoltáicos para electrificación rural. (Training manual: photovoltaics systems for rural electrification) Amazon Cooperation Treaty, Pro Tempore Secretariat. Lima, Perú. 194 pp. 1994. (SPT-TCA/No. 25)

Zonificación ecológica-económica: Instrumento para la conservación y el desarrollo sostenible de los recursos de la Amazonia. Memorias. (Ecological-economic zoning: instrument for the conservation and sustainable development of the resources in the Amazon region. Notes) Amazon Cooperation Treaty, Pro Tempore Secretariat. Lima, Perú. 382 pp. 1994. (SPT-TCA/No. 26)

Plantas medicinales amazónicas: Realidad y perspectivas. (Amazon medicinal plants: reality and perspectives) Amazon Cooperation Treaty, Pro Tempore Secretariat. Lima, Perú. 302 pp. 1995. (SPT-TCA/No. 28)

Propuesta de Tarapoto sobre criterios e indicadores de sostenibilidad del Bosque Amazónico. Memorias. (Tarapoto proposal on criteria and indicators for the sustainability of the Amazon forest. Notes) Amazon Cooperation Treaty, Pro Tempore Secretariat. Lima, Perú. 189 pp. 1995. (SPT-TCA/No. 29)

Memorias de la mesa redonda sobre microempresas agroindustriales como factor de desarrollo sostenible de la región Amazónica. (Proceedings of the Round Table on agroindustrial microenterprises as a factor for the sustainable development of the Amazon region) Amazon Cooperation Treaty, Pro Tempore Secretariat. Lima, Perú. 264 pp. 1995. (SPT-TCA/No. 30)

Biodiversidad y salud en las poblaciones indígenas de la Amazonia. (Biodiversity and health in the indigenous populations of the Amazon region) Amazon Cooperation Treaty, Pro Tempore Secretariat. Lima, Perú. 312 pp. 1995. (SPT-TCA/No. 31)

Inventario y análisis de instituciones y proyectos de conservación y desarrollo en la Amazonia venezolana. (Inventory and analysis of institutions and projects of conservation in the Venezuelan Amazon region) Amazon Cooperation Treaty, Pro Tempore Secretariat. Lima, Perú. 157 pp. 1995. (SPT-TCA/No. 32)

Inventario de proyectos y presencia institucional en la región amazónica colombiana. (Inventory of projects and institutional performance in the Colombian Amazon region) Amazon Cooperation Treaty, Pro Tempore Secretariat. Lima, Perú. 136 pp. 1995. (SPT-TCA/No. 33)

Perspectivas del turismo sostenible en la Amazonia. (Perspectives of sustainable tourism in the Amazon region) Amazon Cooperation Treaty. Pro Tempore Secretariat. Lima, Perú. 125 pp. 1995. (SPT-TCA/No. 34)

Uso y conservación de la fauna silvestre en la Amazonia. (Management and conservation of wildlife in the Amazon region) Amazon Cooperation Treaty. Pro Tempore Secretariat. Lima, Perú. 216 pp. 1995. (SPT-TCA/No. 35)

Base jurídica del Amazon Cooperation Treaty, 4 ed. revisada y actualizada. (Legal framework of the Amazon Cooperation Treaty, 4th edn, revised and updated) Amazon Cooperation Treaty. Pro Tempore Secretariat. Lima, Perú. 300 pp. 1996. (SPT-TCA/No. 36)

Legal framework of the Amazon Cooperation Treaty. 1994–1995. Amazon Cooperation Treaty. Pro Tempore Secretariat. Lima, Perú. 104 pp. 1996. (SPT-TCA/No. 37)

Comisión Especial de Ciencia y Tecnología de la Amazonia (CECTA). Antecedentes constitutivos, actas y anexos de las reuniones. (Special Commission of the Amazon Region on Science and Technology. Background, minutes and anncxcs of the meetings) Amazon Cooperation Treaty. Pro Tempore Secretariat. Lima, Perú. 138 pp. 1996. (SPT-TCA/No. 38)

Comisión Especial de Medio Ambiente de la Amazonia (CEMAA). Antecedentes constitutivos, actas y anexos de las reuniones. (Special Commission of the Amazon Region on the environment. background, minutes and annexes of the meetings) Amazon Cooperation Treaty. Pro Tempore Secretariat. Lima, Perú. 176 pp. 1996. (SPT-TCA/No. 39)

Patentes, propiedad intelectual y biodiversidad amazónica. (Patents, intellectual property and biodiversity in the Amazon Region) Amazon Cooperation Treaty. Pro Tempore Secretariat. Lima, Perú. 456 pp. 1996. (SPT-TCA/No. 40)

Plan de trabajo e informes de actividades de la Pro Tempore Secretariat del Amazon Cooperation Treaty. (Work plan and reports on the administration of the Pro Tempore Secretariat of the Amazon Cooperation Treaty) Amazon Cooperation Treaty. Pro Tempore Secretariat. Lima, Perú. 147 pp. 1996. (SPT-TCA/No. 41)

Desenvolvimento e conservação na Amazônía Brasileira: Inventario e análise de projetos. (Development and conservation in the Brazilian Amazon region: inventory and analysis of projects) Amazon Cooperation Treaty. Pro Tempore Secretariat. Lima, Perú. 247 pp. 1996. (SPT-TCA/No. 42)

Cultivo del pijuayo (Bactris gasipaes Kunth) para palmito en la Amazonia. (Cultivating pijuayo for palmito in the Amazon region) Amazon Cooperation Treaty. Pro Tempore Secretariat. Lima, Perú. 153 pp. 1996. (SPT-TCA/No. 43)

Frutales y hortalizas promisorios de la Amazonia. (Promising fruits and plants in the Amazon Region) Amazon Cooperation Treaty. Pro Tempore Secretariat. Lima, Perú. 450 pp. 1996. (SPT-TCA/No. 44)

Inventory of institutions and projects operating in the Amazon region of Suriname. Amazon Cooperation Treaty. Pro Tempore Secretariat. Lima, Perú. 106 pp. 1996. (SPT-TCA/No. 45)

El cultivo del camu camu (Myrciaria dubia H.B.K. Mc Vaugh) en la Amazonia peruana. (Cultivation of Camu-Camu in the Peruvian Amazon Region) Amazon Cooperation Treaty. Pro Tempore Secretariat. Lima, Perú. 95 pp. 1996. (SPT-TCA/No. 46)

Piscicultura amazónica con especies nativas. (Amazon fishing with native species) Amazon Cooperation Treaty. Pro Tempore Secretariat. Lima, Perú. 169 pp. 1996. (SPT-TCA/No. 47)

Crianza familiar del majaz o paca (Agouti paca) en la Amazonia. (Smallholdings of majaz or paca in the Amazon region) Amazon Cooperation Treaty. Pro Tempore Secretariat. Lima, Perú. 43 pp. 1996. (SPT-TCA/No. 48)

Comisión Especial de Transportes, Comunicaciones e Infraestructura de la Amazonia (CETICAM). Antecedentes constitutivos, actas y anexos de las reuniones. (Special Commission of the Amazon Region on Transport, Communications and Infrastructure. Background, minutes and annexes of the meetings) Amazon Cooperation Treaty. Pro Tempore Secretariat. Lima, Perú. 299 pp. 1996. (SPT-TCA/No. 49)

Proposal of criteria and indicators for sustainability of the Amazon forest. Results of the regional workshop. Amazon Cooperation Treaty. Pro Tempore Secretariat. Lima, Peru. 149 pp. 1995. Publicación trilingüe: inglés, español y portugués. (SPT-TCA/s.n.)

Forthcoming

Tierras y Areas Indígenas en la Amazonia: Una experiencia regional participativa. (Indigenous lands and areas in the Amazon region: a regional participatory experience). 200 pp. approx.

"Estrategias y Acciones para un Programa Regional de Promoción de la Producción Sostenible y Utilización de Frutales y Hortalizas" 200 pp. Memorias de la Mesa Redonda sobre la Complementariedad de la Producción Frutihortícola Amazónica con el Desarrollo de Microempresas Agroindustriales en los Países del TCA. (Strategies and actions regarding a regional programme for the promotion of sustainable production and use of fruit trees and vegetables. Minutes of the Round Table on Complementarity between Amazon Fruit and Horticultural Production and Agro-industrial Microenterprise Development in the Countries of the ACT).

"Manual de Cultivo y Uso de Plantas Medicinales en la Amazonia" (Manual of breeding and use of medicinal plants in the Amazon region).

"Procesamiento a Pequeña Escala de Frutas y Hortalizas Amazónicas Nativas e Introducidas – Manual Técnico". 140 pp. approx. (Small-scale processing of native and imported fruits and vegetables of the Amazon region – technical manual).

In Preparation

Comisión Especial de Turismo de la Amazonia (CETURA). Antecedentes constitutivos, actas y anexos de las reuniones. 250 pp. approx. (Special Commission of the Amazon Region on Tourism (CETURA). Background, minutes and appendixes of the meetings).

Comisión Especial de Salud de la Amazonia (CESAM). Antecedentes constitutivos, actas y anexos de las reuniones. 250 pp. approx. (Special Commission of the Amazon Region on Health (CESAM). Background, minutes and appendixes of the meetings).

Comisión Especial de Asuntos Indígenas de la Amazonía (CEAIA). Antecedentes constitutivos, actas y anexos de las reuniones. 350 pp. (Special Commission of the Amazon Region on Indigenous Affairs (CEAIA). Background, minutes and appendixes of the meetings).

Memorias del Seminario-Taller "Propuesta Metodólogica para la Zonificación Ecológica-Económica para la Amazonia". 500 pp. approx. (Proceedings of the

Seminar Workshop "Metodological Proposal for the Participatory Ecological-Economical Zoning for the Amazon Region."

Inventario de Proyectos y Presencia Institucional en la Región Amazónica Peruana. 80 pp. approx (Inventory of projects and institutional performance in the Peruvian Amazon region).

Informe de Actividades y de Gestión de la Pro Tempore Secretariat. Perú. SPT-TCA. 100 pp. approx. (Reports on the activities and administration of the Pro Tempore Secretariat of the Amazon Cooperation Treaty).

Video

Manual de Procesamiento de Frutas y Hortalizas (Manual of fruit and vegetable processing).

CD-ROM

Bases de Datos sobre Poblaciones Indígenas de la Amazonia (Database on indigenous populations in the Amazon region).

Base de Datos sobre Poblaciones Indígenas de la Región Amazónica Peruana (Database on indigenous populations in the Peruvian Amazon region).

Colección de publicaciones de la Pro Tempore Secretariat (1994–1997) (Compilation of publications edited by the Pro Tempore Secretariat of the Amazon Cooperation Treaty 1994–1997).

Part 2
The Plata river basin

4

The hydrology of the Upper Paraguay basin

Carlos E.M. Tucci, Fernando Genz, and Robin T. Clarke

Introduction

The hydrology of the Upper Paraguay river basin is a major factor influencing regional environment. In the past, marked spatial and temporal variability in rainfall, and consequently in runoff and evaporation losses, have resulted in distinct changes to the drainage system and to the region's fragile environment, and proposals for future development will, if implemented, have very substantial and not wholly desirable consequences.

The basin of the Upper Paraguay comprises two distinct areas with differing hydrological behaviour: namely, the Planalto and the Pantanal, and the differences in the hydrology of the two regions strongly influence both terrestrial and aquatic ecosystems and human activity. The Planalto consists broadly of land lying above the 200 m contour; its annual rainfall exceeds 1,400 mm with distinct seasonal distribution and rapid drainage. The land is used principally for agriculture and cattle, and since the 1970s the area planted to soya has grown rapidly. The Pantanal, on the other hand, is one of the world's largest wetlands, with area about 124,000 km^2 lying below the 100 m contour. Annual rainfall is less than potential evaporation, drainage is very slow because of shallow gradients, and sediment deposition is extensive. The ecosystem of the Pantanal is very distinctive and its preservation is important for Brazil. Nevertheless, current human activities and proposed developments constitute a considerable threat to its existence. Although cattle production in the Pantanal is widespread, there is increasing mineral development; intensive soya production in the Planalto is associated with increased soil erosion and increased sediment deposition in the Pantanal; the unique Pantanal ecosystem is being exploited for the development of tourism; but most important of all, there are proposals to convert the Paraguay river into a navigable waterway to export the region's agricultural products and for trade.

The consequences of these developments, both actual and proposed, must be carefully studied if the Pantanal is to avoid total destruction, and to establish how development will influence the region it is essential to understand its hydrology. This paper discusses the principal elements of the region's hydrological behaviour, and how its hydrology influences the environment and human activity within it.

Characteristics of the Upper Paraguay

The basin of the Paraguay river includes parts of four countries (Brazil, Bolívia, Paraguay, and Argentina) and is one of the most important in South America. The Paraguay joins the river Paraná near Corrientes in Argentina to become the main river of the Plata drainage basin. Figure 1 shows its position in the South America subcontinent and figure 2 its location in the Plata river basin.

The Paraguay rises in Brazil near the divide to the south of the Amazon drainage basin, and its boundaries are uplands ranging in height from 500 m to 1,400 m. The Upper Paraguay basin is conventionally taken as the area lying upstream of the Porto Esperança flow gauge, and it extends to about 360,000 km². Figure 3 shows the Planalto and Pantanal subdivisions of the Upper Paraguay.

The basin has two main points where flow is constricted (see fig. 3) giving rise to what are in essence large natural reservoirs. One of these controls is near São Francisco and the other is upstream of Porto Murtinho; at both points of constriction, rock formations limit the flow section.

Mean rainfall, evapotranspiration, and flow

Figure 4 shows that the mean annual rainfall in the Pantanal is of the order of 1,180 mm, while potential evapotranspiration is about

Fig. 1 **Location of Upper Paraguay basin in South America**

1,370 mm. Figure 5 illustrates the variability in rainfall and evapotranspiration in the basin of the River Taquari, lying within the Pantanal. Mean annual temperature is 25°C, with a mean minimum of 20°C and maximum of 32°C. Figure 6 shows that specific flows in the Planalto vary between 15–20 $l/s^{-1}.km^{-2}$, falling off significantly downstream. Mean outflow at Porto Esperança over a 12-year period (1970–1981) was 2,165 $m^3.s^{-1}$, whilst the total of all streams entering the Pantanal from the Planalto was 2,058 $m^3.s^{-1}$. These figures show a mean flow from the Pantanal of 107 $m^3.s^{-1}$, giving a specific flow of only 0.91 $l/s^{-1}.km^{-2}$. Inflows to the Pantanal and outflows from it are therefore almost in balance.

105

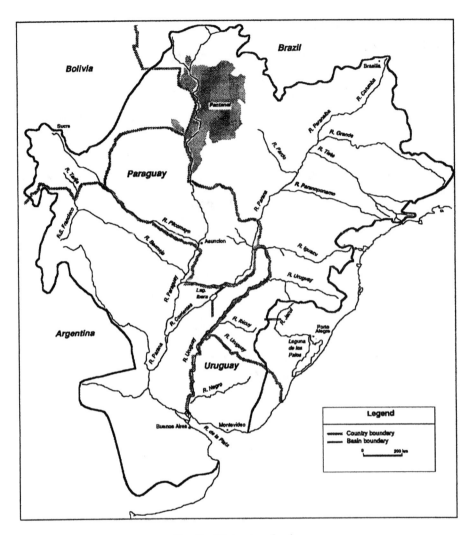

Fig. 2 **Plata river basin**

In table 1, Qe is the annual total inflow to the Pantanal and Qs the annual outflow at Porto Esperança; P and Er and mean areal rainfall and estimated actual evaporation respectively; and $D = Qe + P - Er - Qs$ for each of the years 1970–1981. The final column D shows in which years the balance was negative or positive; averaged over all 12 years, which were above average in terms of rainfall, inflow exceeded outflow and the flooded area increased from 6,770 km^2 in 1970 to 52,697 km^2 in 1981 (estimates given by Hamilton et al., 1996).

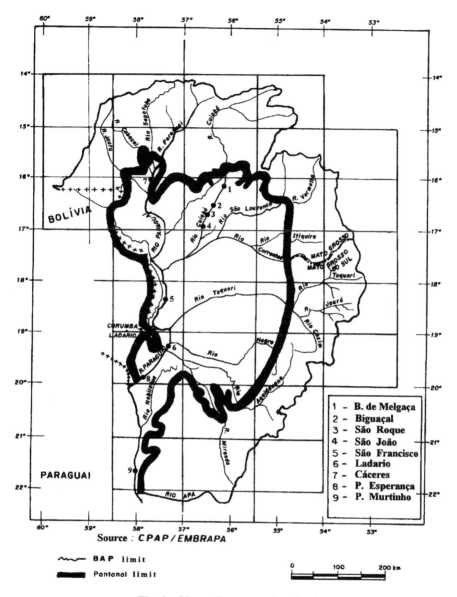

Source : *CPAP / EMBRAPA*

~~~ BAP limit

▬▬ Pantanal limit

0   100   200 km

Fig. 3   **Upper Paraguay river basin**

Behaviour of runoff as it crosses the Pantanal

The drainage channels leaving the Planalto are in the form of a fan converging on the Pantanal, with various rivers crossing it to form the Paraguay river (fig. 3). This river runs along the western edge of

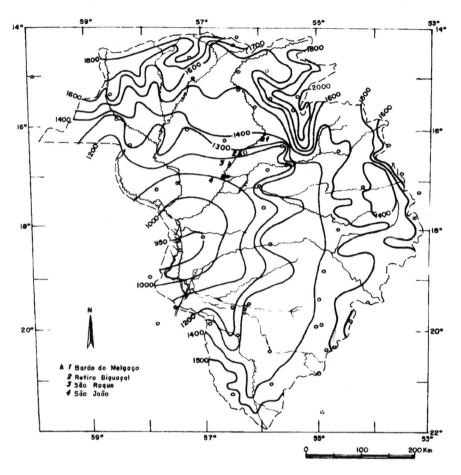

Fig. 4 **Rainfall distribution**

the basin, and its principal tributaries which cross the Pantanal are the Cuiabá, São Lourenço, Piquiri, Taquari, Negro, and Miranda rivers.

Flow in these rivers is drastically reduced where they enter the Pantanal because of the abrupt decrease in gradient, with a consequent deposition of sediment on the river bed and reduction in erosive power causing a reduction in channel cross-section relative to that farther upstream. In periods of flood, downstream sections of the Pantanal rivers have smaller conveyance than those upstream, resulting in extensive overbank spillage to a broader channel, and the greater the flood the more extensive is the overbank spillage. However the plain of the Pantanal contains a large number of depressions which are filled during floods, forming a landscape filled with small

108

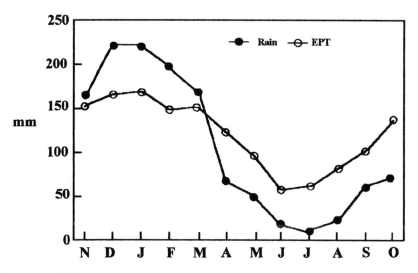

Fig. 5 **Rainfall and evapotranspiration variation in Pantanal (Hamilton et al., 1995)**

lakes which link together as the water level rises and retain water as the level falls in the main channel. A large part of the inflow from upstream is therefore retained as storage in depressions having no direct link with the main channels of rivers flowing acoss the Pantanal.

In periods when the rivers are contained within the banks of the main channel and there is no rainfall, the volume of water retained in the depressions is reduced by evaporation losses and by infiltration to groundwater; but fine sediments and organic material carried from upstream are deposited on the beds of the depressions so that infiltration to groundwater is limited. In addition, the Pantanal experiences high temperatures almost throughout the entire year so that evaporation losses are high.

The mechanisms described above are the cause of the reduction in rates of transport of water and sediment with distance downstream (IPH, 1983). Tucci and Genz (1996) quantified the processes involved through study of various reaches of the Paraguay river within the Pantanal, one of which was on the Cuiabá river between Barão de Melgaço and São João. This reach has two additional gauging sites, namely São Roque and Retiro Biguaçal. Table 2, giving mean annual flows for the Cuiabá river in those years for which all stations have complete records, shows the reduction in mean flow that occurs with distance downstream. Table 3 gives the 30-day low flows in years with complete records, and it is seen that the 30-day low flow increases

109

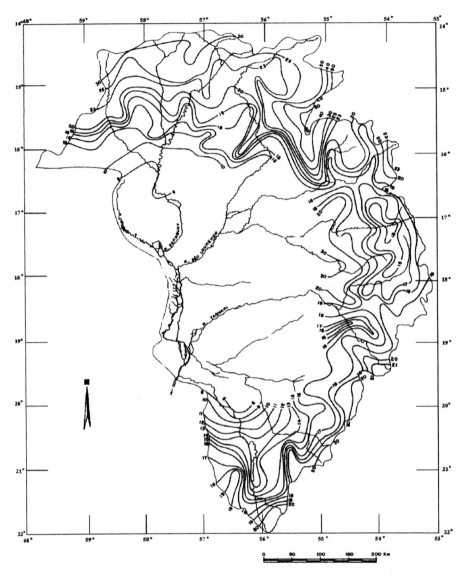

Fig. 6 **Specific discharge of Upper Paraguay river basin**

with distance downstream, particularly in the reach between Barão de Melgaço and Biguaçal. These low flows represent within-bank flow, and show a picture that is the converse of what is seen with mean flows, which are influenced by overbank spillages lost to evaporation.

Figures 7 and 8 show the reduction in flood volume for the year

Table 1   **Water balance in Pantanal area**

Year	Qe km³	Qs km³	P km³	E km³	D km³
1970	40.509	41.060	122.141	129.21	−7.62
1971	35.253	38.095	144.724	129.21	12.67
1972	44.829	40.522	134.444	129.21	9.54
1973	49.231	44.182	133.075	132.22	5.90
1974	71.748	83.823	157.175	141.19	3.91
1975	62.98	65.090	134.444	139.61	−7.28
1976	69.939	74.961	164.949	139.60	20.33
1977	76.016	83.129	175.574	146.28	22.18
1978	77.818	73.100	134.383	144.570	−5.47
1979	87.056	98.045	159.470	142.85	5.63
1980	86.494	91.171	155.607	142.54	8.39
1981	77.014	85.967	136.234	144.95	−17.67
Mean	64.91	68.26	146.02	138.63	

Table 2   **Mean annual flows ($m^3.s^{-1}$) at stations on the Cuiabá river**

Year	Barão de Melgaço (1)	Biguaçal (2)	São Roque (3)	São João (4)
1971	224	206	202	197
1972	286	238	291	207
1973	323	280	269	232
1974	457	343	328	265
1975	384	324	311	250
1976	366	328	316	261

Note: ( ) Stations numbered from upstream to downstream.

1994. The reductions in volumes shown by the hydrographs are significant in all reaches. For the reach of the Cuiabá river between Barão de Melgaço and São João, flows were analysed for the period March–April of 1974 (figs. 7 and 8) and a 62 per cent reduction in volume of flow was calculated. In periods of flood when the river overspills, flood volumes therefore decrease with distance downstream. During the dry season when the river flows within its banks, flow increases with distance downstream since there are no losses by evaporation from flooded areas and gains from subsurface flow, although the gains are minor in some reaches.

Table 3 **Mean 7-day low flows (m³.s⁻¹)**

	Barão de Melgaço (1)	Biguaçal (2)	São Roque (3)	São João (4)
1971	67.9	79.0	78.5	76.0
1972	79.1	88.2	89.1	89.1
1973	84.9	91.6	90.7	90.5
1974	113.4	122.4	120.6	125.9
1975	94.4	109.1	107.5	99.5
1976	98.4	118.1	117.2	112.4
1977	119.3	130.7	132.3	124.6
1978	95.5	120.7	130.4	116.7
1979	125.5	143.3	151.6	152.4
1980	120.8	141.9	151.2	145.7

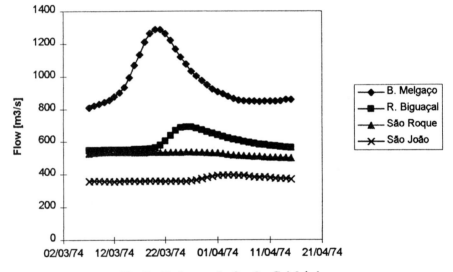

Fig. 7 **Hydrographs for the Cuiabá river**

Figures 7 and 8 clearly show the reservoir effect causing attenuation of hydrographs where flows leave the Planalto for the Pantanal flood plain. Thus the hydrological evidence suggests that if the conveyances of rivers crossing the Pantanal were to be improved, mean flows could very well increase with distance downstream, but at the cost of reduction in flooded areas with consequent modification to the ecosystems that they support.

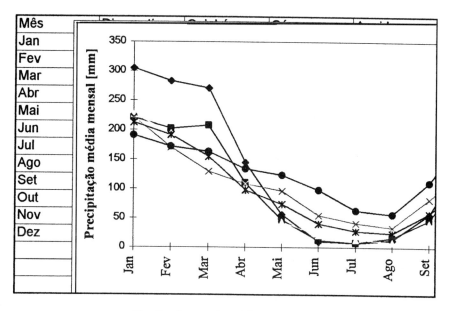

Fig. 8   **Accumulated flow – March–April 1974**

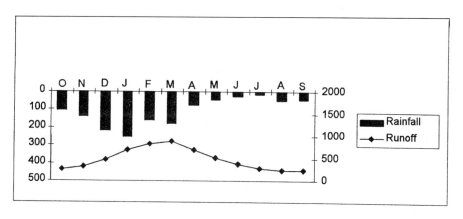

Fig. 9   **Mean monthly rainfall and flow: Paraguay river at Cáceres**

Temporal and spatial distribution of Pantanal flooding

Analysis of rainfall data from the Upper Paraguay shows that the wet season usually extends from October to April, with some variation in the date of onset. In the upper part of the basin, the wet season tends to fall between October and March (figure 9). However, there is also some evidence of an east-west trend in the date of wet-season onset,

113

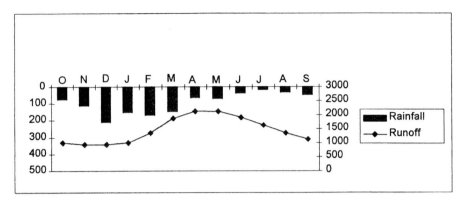

Fig. 10   **Mean monthly rainfall and flow: Paraguay river at São Francisco**

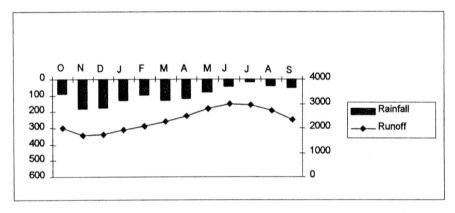

Fig. 11   **Mean monthly rainfall and flow: Paraguay river at Porto Murtinho**

which is accompanied by greater variability in rainfall depth.

Figures 9 and 10 show the drift in the months of greatest rainfall and mean flow – from Cáceres to São Francisco, for example – with distance downstream. At Cáceres, mean flow is greatest in March, at the end of the wet season; at São Francisco on the other hand, mean flow is greatest in April and May and smallest in the wet months of December, January, and February. At Porto Murtinho (figure 11), flow is greatest in June and July, completely out of phase with the upstream rainfall regime. The pattern observed at Cáceres is also repeated in other tributaries of the Paraguay, such as the Cuiabá river.

These figures show that flow across the Pantanal flood plain is very slow, because of the small conveyance within the channel of the River Paraguay and its tributaries, thereby creating extensive spillage (see below).

In the Planalto, rivers are full or overflowing during the wet season of October to March, and enter recession during the period of low rainfall from April to September, when flow is maintained by discharge from aquifers and return flow from flooded areas. In the area between the Planalto and where the rivers draining it enter the Paraguay river, spillage to the flood plain occurs at the time of the wet season, because of local precipitation and the drainage from the Planalto. At this time, the Paraguay river proper has not yet risen sufficiently or is still in recession, being out of phase by two or three months. As the level of the Paraguay is lower, flows in its tributaries can enter it more easily without giving rise to impoundment and consequent backwater effects.

When the tributaries move into recession, the levels in the Paraguay river begin to rise, spilling on to its flood plain, damming its tributaries and maintaining the flood in their areas of influence for a large part of the year.

The most important conclusions to be drawn involve the following:
- The Paraguay river exerts an influence on a part of each of each of its tributaries, causing prolonged flooding. Between October and March, the flooding is caused by flows in the tributaries themselves, but for the rest of the year the flood originates from the Paraguay river itself. The extent of the Paraguay's influence depends principally on within-year conditions, on tributary channel conditions, and on regional topography. In areas nearer to the Paraná river, the two floods may occur together because the seasonality of rainfall is less marked. Flood levels can then be high, even though the wet season is not particularly extreme.
- Away from the areas influenced by the Paraguay river, hydrological behaviour of the Pantanal bears a strong seasonal relationship to the wet season occurrence, and to the dry period between April and September.
- Because of silting within tributary channels when wet seasons are less marked, and the greater width of flow during seasons that are very wet, both the flood level in the Paraguay river and the spatial extent of its area of influence may vary appreciably from year to year. This can be seen particularly in the sequence of very dry years from 1960–1972, and of wet years following 1973.

115

How representative are the hydrological series?

The hydrological records of the Upper Paraguay basin are limited principally in their temporal representativity. This is especially true of the flow records, since only one station – Ladário – has a long record of water level, beginning in 1900. All other water-level records began in the 1960s.

Rainfall records

There are six sites with long rainfall records (from 47 to 88 years) but the remaining records all date from the late 1960s. An earlier study (IPH, 1983) analysed the mean and standard deviation of annual rainfall, and of wet-season rainfall, for the sites established in the 1960s, so the present study used data from the six sites with longer records mentioned above. The records from these sites were divided into two halves and statistics calculated from the two halves were tested for homogeneity. The tests showed that neither the means nor the variances of annual rainfall differed significantly at any of the six sites; wet-season rainfall in the two periods of record differed significantly at one site only, but the difference was only just significant at the 5 per cent level.

The records show that on average 85 per cent of the rain falls in the wet season. The proportion is higher at sites in the north of the basin, and is less for sites in the east and south where several consecutive months may have little or no rain. Figure 12 shows the variability of monthly rainfall at the sites used in the analysis; seasonality of rainfall distribution is more marked in the northern sites than in those to the south or east.

Records of flow and water level

Analysis of the runoff water level records encounters the following problems:
(i)   only one site exists with records beginning before 1960;
(ii) changes in the river bed cast doubt on the validity of water level records and on rating curves;
(iii) DNOS (the flood-control organization which maintained the sites in the network) ceased operation in 1990. Some, but not all, sites were reactivated, but the selection of sites to be reactivated was determined by considerations of cost and not by hydrological need;

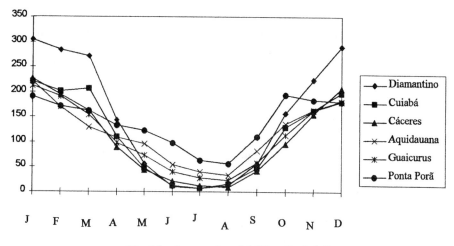

Fig. 12   **Seasonal variability of rainfall**

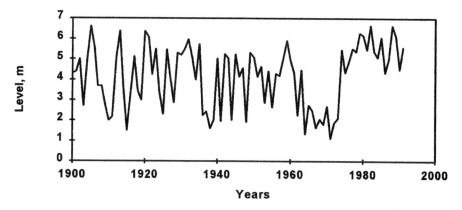

Fig. 13   **Annual maximum water level at Ladário, Paraguay river (1900–1992)**

(iv) the majority of sites have no measurements of discharge in the upper parts of their rating curves.

Comparisons with the record from Ladário (1900–93) provide the means to determine whether the records initiated in the 1960s are representative of the longer period. Figure 13 gives the annual maximum water levels at Ladário for the whole period of record; between 1900 and 1960 the annual maxima fluctuate around 4.0 m, whilst between 1960 and 1973 they fluctuate around 2.0 m. Subsequently, between 1970 and 1991 the annual maxima fluctuate around 5.0 m. These variations are very large for a region where most of the river beds are alluvial and which (Leopold et al., 1964) are formed by

117

floods typically with return period 1.5 years. Therefore in the period 1960–73 reductions in the energy of flow and consequent siltation caused rivers to form beds corresponding to flow capacities lower than those of the preceding period. The inhabitants of the region – especially cattle ranchers – began to make use of areas which had previously been flooded for long periods. When, subsequently, annual maximum levels increased to fluctuate around a mean level more than twice that of the previous period, areas which had been free of flooding for several months of the year became almost permanently flooded.

The main questions suggested by these records are as follows: (i) can the observed variation in annual maximum water levels be explained by changes to the river bed at Ladário? (ii) is it possible that the measuring scale at Ladário was moved without the change being recorded? (iii) how far can the changes in annual maximum water level be explained by land-use changed and other human activity? (iv) can the fluctuations in water level be explained in terms of climatic variation?

Since there are no records of length comparable with that at Ladário, other records of water level and flow were studied to see whether the behaviour of these variables corresponded with that shown by the Ladário record. Figure 14 shows data from other sites

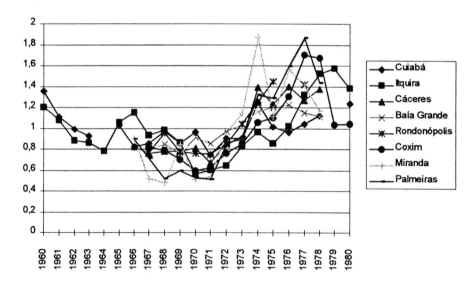

Fig. 14 **Adimensional flows at sites on the Upper Paraguay basin**

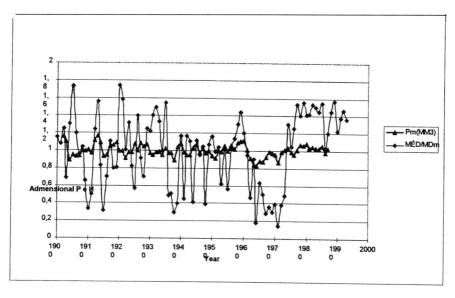

Fig. 15 **Running mean of rainfall at Cuiabá and water levels at Ladário (adimensional)**

in non-dimensional form, and it is seen that there is a large variation in the 1960s and 1970s. This suggests that the changes to the river-bed at Ladário, if they occurred, do not entirely explain the wide fluctuations in annual maximum water level observed at that site.

To answer the remaining questions listed above, variations in water levels at Ladário were compared with rainfall records in the contributing basin. Figure 15 shows mean water levels at Ladário and a three-day moving average of rainfall at Cuiabá for the available period. The figure shows that the rainfall follows the same pattern, albeit with less variability, during the 1960s and after the year 1973. The difference in the extent of the variability is mainly due to the differences in magnitude of the two variables which result in differences in amplitude.

Analysis of the available records does not disprove the hypothesis that the pattern in annual maximum flow at Ladário is a consequence of climatic variation. This explanation gains some qualitative support from records of other Brazilian drainage basins, although further study, particularly of low flow conditions, is necessary since these depend much more upon shortfall in rainfall, particularly in basins where infiltration is low.

119

Sediment yield and transport

The production and transport of sediment in the Upper Paraguay basin is of great concern because of the environmental impact of human activities and economic development of the region. In the Planalto, there has been a dramatic increase since the decade of the 1970s in areas planted to annual crops, principally soya, and this has significantly increased soil erosion and the sediment transported into the Pantanal. At the same time, the short-term increases of annual rainfall in the upper part of the basin have caused soil loss in the Planalto with deposition in some reaches and, in the Pantanal, greater deposition of sediment and reduced channel conveyance.

Within the Pantanal itself, land-use changes have been largely restricted by the hydrological regime, although some dykes have been constructed to control flooding. However during floods, rivers may change course by cutting across meanders, and since a land-owner then loses part of his property, he may seal off the new course, so that the river will revert to its old one; where this occurs, there is a consequent high mortality of fish and other aquatic life in the cut-off section. Borges et al. (1996) describe the reduced concentration of suspended sediment in rivers leaving the Planalto and crossing the Pantanal, from a maximum of 1 $d^{-1}km^{-2}$ near the headwaters of the São Lourenço river and falling to 0.2 t $d^{-1}km^{-2}$ in the Taquarí river near its junction with the Paraguay river, showing a heavy deposition of sediment within the Pantanal. The same authors estimate a reduction of 50 per cent in bed-load material where the left-bank tributaries enter the Paraguay; along the Paraguay itself the reduction is estimated at 20–30 per cent.

Water quality

The basin has a network of water quality stations which is being extended to allow estimation of pollutant loads. Pollution from domestic and industrial wastes is as yet of only local occurrence and levels of dissolved oxygen (DO) are satisfactory over most of the basin although rivers receive untreated domestic sewage. Since flow is generally high, population densities low and industrial development not yet widespread, pollution impacts are small. Significant reduction in DO, when this occurs, is often the result of natural factors acting in the river system, an DO levels can sometimes fall to near zero:

- During floods, the rivers tend to erode their beds causing a resuspension of benthic material; Borges et al. (1996) mention that during their data collection, samplers often became filled with dead leaves, the decomposition of which makes heavy demands on dissolved oxygen.
- Particularly at the beginning of the flood season, there tends to be a major influx of new vegetable matter into the river system, which requires much oxygen for its decomposition.

The decomposition of this vegetable matter can cause such a heavy oxygen demand locally, that fish die in the de-oxygenated water. Other potentially polluting factors, such as mineral extraction, pesticide use and contamination of recharge areas, are being monitored and controlled by environmental agencies.

Conclusions

The hydrological behaviour of the Upper Paraguay basin, and of the Pantanal in particular, exerts a strong influence upon how its unique wetland environment can be preserved whilst allowing for a measure of sustainable development.

Rainfall in the Pantanal is less than potential evaporation, so that the area has a potential deficit; areas are flooded by runoff from the Planalto. The Pantanal, like other wetlands, falls under threat when the drainage system increases its transport capacity, which can occur because of channel dredging, removal of obstructions to flow, and the construction of dykes. In broad terms, increasing the transport capacity of the Upper Paraguay would reduce the volume of water retained in the Pantanal, possible changing it from wetland to serrado, but more research is needed before firm conclusions can be drawn concerning how and where the changes would occur; it is possible that minor localized changes may have limited local effects, whilst the combined effect of many such localized changes may affect the character of the Pantanal very substantially.

On the other hand, the livelihoods of the Pantanal inhabitants have been strongly influenced by the climatic variations described in this paper. Ranches which were free of flooding for much of the year (7–9 months) in the 1960s have seen the flood-free period reduced to only 3–5 months in more recent decades. This has created pressure from some landowners for the construction of dykes, some of which have been built. Nevertheless, some questions remain unanswered: was the period of the 1960s, when flooded areas were substantially smaller and the duration of flooding less, an anomaly? Is the subsequent picture of the last 25 years, with high flood levels not only in the Upper Paraguay but also in the Paraná river, likely to be maintained?

For people to live with the possible changes, whether from natural climate variation or modified land-use and other kinds of economic development, it will be necessary to define which zones, in both urban and rural areas, will be subject to flooding with any given frequency. Schemes for forecasting both the time and extent of flooding will be needed so that inhabitants can take appropriate action to defend property and livelihood; and since the onset of the annual flood is so gradual, it should be possible for warnings to be given well in advance – allowing, for example, cattle to be removed from low-lying areas and provision made for their food supply.

References

Borges, A., Semmelman, F., Bordas, M., and Simões Lopes, M. 1996. "Fluvio-morfologia." In: *Hidrossedimentologia do Alto Paraguay*. Instituto de Pesquisas Hidraulicas, UFRGS: Fundação do Meio Ambiente do Mato Grosso (FEMA/MT), Secretaria do Meio Ambiente do Mato Grosso do Sul, Ministério do Meio Ambiente (SEMA).

Hamilton, S.K., Souza, O.C., and Countinho, M.E. 1995. "Dynamics of floodplain inundation in the alluvial fan of the Taquari River (Pantanal, Brazil)." Proceedings of the International Society of Theoretical and Applied Limnology, 1995 Congress. Brazil: São Paulo.

Hamilton, S.K., Sippel, S.J., and Melack, J.M. 1996. "Inundation patterns in the Pantanal wetland of South America determined from passive microwave remote sensing." *Hydrobiologie*, January.

IPH. 1983. *Regionalização das Vazões do Alto Paraguai*. Instituto de Pesquisas Hidráulicas/Eletrobrás.

Leopold, L.B., Wolman, M.G., Miller, J.P. 1964. *Fluvial Processes in Geomorphology of San Francisco*. W.H. Freeman and Co.

Tucci, C.E.M., and Genz, F. 1996. "Comportamento Hidrológico." In: *Hidrossedimentologia do Alto Rio Paraguay*. Instituto de Pesquisas Hidraulicas, UFRGS; FEMA/MT; SEMA.

Acknowledgement

This study was developed as part of PCBAP – Conservation Plan of Upper Paraguay Basin, funded by the Environment and Water Resources Ministry of Brazil, Mato Grosso and Mato Grosso do Sul States.

5

Water-resources management of the Plata basin

Victor Pochat

Introduction

The Plata basin – due to its extension, the variety of its geography and its climatology, the discharge of its rivers, the magnitude of its population, its economic importance, its social problems, the history of its development and its potentiality – is one of the international basins in the world worthy of in-depth study. Its institutional organization has been discussed by several authors (Cano, 1979; Barberis, 1988; del Castillo de Laborde, 1991). However, there is an aspect that requires to be analysed in greater detail, since it refers to a practical experience which, with its accomplishments and its failings, may be adaptable to other similar cases. It deals with the role played by the experts related to water resources within the institutional system of the basin.

This paper tries to show how their activities evolved, by dividing the almost 30 years of performance of the system into stages, each with its distinctive characteristics.

Organization 1967–1972: Differences in criteria and groups of experts

The first meeting of the Foreign Affairs Ministers of the five countries comprising the Plata basin – Argentina, Bolivia, Brazil, Paraguay, and Uruguay – was held in Buenos Aires in February 1967.

As a result of that meeting, the Ministers issued a declaration, saying "that it is a decision of our governments to carry out the joint and integral study of the Del Plata Basin, with a view to the realization of a programme of multinational, bilateral and national works, useful for the progress of the region."

As a first step they created the Intergovernmental Coordinating Committee of the Countries of the Plata Basin (CIC), with the aim of drawing up a statute for its definitive constitution. Since the CIC is an organization run by diplomats, the declaration proposed each country would appoint their own technical advisers. Further, the declaration ruled that, to achieve the objective of the integral development of the basin, that study should take into account – in relation to water resources – in the main the following subjects: facilities and assistance to navigation; establishment of new fluvial ports and the improvement to existing ones; hydroelectric studies with a view to energy integration of the basin; installation of water services for domestic, sanitary, and industrial uses, and for irrigation; floods or inundations and erosion control, and the conservation of animal and vegetal life.

During the second meeting of Ministers, held in Santa Cruz de la Sierra in May 1968, the Statute of the CIC was approved. It was entrusted to draw up a treaty in order to enforce the institutionalization of the basin. At the same time, it was agreed to carry out preliminary studies in relation to concrete projects presented by the member countries. Among the projects shared by the five member countries, the following should be noted:

A-1 Construction of a port in Bolivian territory on the Paraguay river and its connection to the railroad network (Bush port).

A-2 Hydrometeorology and future establishment and performance of the regional network of hydro-meteorological stations.

A-3 Inventory and analysis of basic information on the basin's natural resources and related subjects.

A-4 Study of problems to be solved and projects of measures to be taken (dredging, obstacle removal, signalling, buoyage, etc.) in order to allow permanent navigation and to secure its maintenance in the Paraguay, Paraná, Uruguay, and Plata rivers.

...

A-7 Assessment of the ichthyologic resources of the basin.

In relation to the specific projects presented by member countries, should be mentioned:

B-1 Regulation of the Bermejo and Pilcomayo rivers from their headwaters.

B-2 Study for the modernization of the Buenos Aires and Montevideo ports.

B-3 Adaptation and settlement of Asunción port.

B-4 Studies for the modernization and possible integration to the system of the Río Grande port.

B-5 Study of the Santa Lucía river basin.

B-6 New updating of information about the specific project and completion of work at Salto Grande.

The chancellors also recommended that the CIC studied and drew up a statute to fit the use and administration of water resources in relation to the rital purpose of integral and harmonious development of the Plata basin expressed at the first meeting of Ministers.

On 23 April 1969, during their first extraordinary meeting held in Brasília, the Ministers signed the Treaty of the Plata Basin which, in its Article 1, only paragraph, says:

The Contracting Parties agreed to unite efforts with the objective of promoting the harmonious development and the physical integration of the Del Plata basin and of its area with direct and considered influence.

Single paragraph: With that purpose, they will promote within the ambit of the basin, the identification of areas of common interest and the promotion of research, programmes and works, as well as the formulation of operative agreements and juridical instruments they consider necessary and that tend to:

(a) give facilitation and assistance as regards navigation;

(b) promote reasonable utilization of water resources, especially by means of the regulation of watercourses and their multiple and equitable development;

(c) achieve the preservation and the improvement of animal and vegetal life;

...

(h) promote other projects of common interest and especially those that have relation to the inventory, assessment and development of the natural resources of the area;

(i) further understanding of the Plata basin.

During that first extraordinary meeting and the simultaneous third ordinary meeting in April 1969, the Ministers of Foreign Affairs recommended to the CIC that they should constitute a Group of Experts in order to "consider extensively the subject of water resource" and establishing "that the respective group of experts should present its report as soon as possible, by considering the importance and complexity of the issue."

This group of experts in charge of studying the water resource met in Rio do Janeiro from 5 to 9 August 1969. However, it was not possible for them to make progress as there was no consensus in relation to the working agenda itself. The representatives of Argentina and Uruguay insisted on an analysis of a preliminary statute presented to the CIC by the Uruguayan delegation in December 1968 – following the recommendation of Santa Cruz de la Sierra – while the delegations from Bolivia, Brazil, and Paraguay preferred to explore, within the so-called inductive method, the wide areas on which there was unanimous consensus, interpreting the spirit and the word of their original mandate. The group then applied to the CIC, requesting an explanation about "the scope of its mandate, within the parameters of the Brasília Act."

This brought out the different approaches of the various countries about the juridical concepts to be applied in relation to the development of the basin rivers. Particularly, the delegates from Argentina and Brazil kept up a valuable interchange of opinion, by means of speeches in committee meetings[1] as well as by presentation of notes,[2] between March and April 1970 (COMIP, 1992).

In order to respond to the request of the group of experts, the Committee held a special meeting for delegates to air their respective points of view. In addition, the Brazilian delegation continued with bilateral talks with the heads of the other delegations.

This allowed for the Uruguayan representative to present a provisional agenda for the second meeting of the group of experts. Such a document suggested, as the main mandate for the group, studying "widely and in a simultaneous manner, all the aspects related with the water resource, especially by means of the regulation of the watercourses and its multiple and equitable development." With that aim, the group of experts should "identify the areas of common interest, to further research, programmes and works as well as the formulation of operative agreements or juridical instruments they consider necessary." After establishing the purpose of the tasks and the methodology to be used, the agenda specified four basic subjects to be studied: (a) exchange of hydrological and meteorological data; (b) pollution; (c) evaluation of the influence of the hydrological aspects; (d) national or multinational development of the basin waters.

The second meeting of the group of experts for water resources was held in Brasília, between 18 and 22 May 1970, reaching interesting conclusions, presented as suggestions to the CIC. As this was an initial and comprehensive meeting, it is worth repeating some part of it (COMIP, 1992):

The Group of Experts, considering:
that within the framework of the development of the States of the Plata Basin, it is convenient to proceed to the exchange of hydrological and meteorological data, being necessary to standardize the techniques of observation, installation and enlargement of stations and taking into account the recommendations of OAS, the Hydrological International Decade and WMO,

suggests to the Intergovernmental Coordinating Council to recommend to the States of the Basin:
1. To enlarge and improve their basic networks ...
2. To proceed in the following manner for the observation and exchange of hydrometeorological data:
    2.1. The installation of a main climatological station at the centre of gravity of each climatic zone, without considering its present economic importance and looking to the future development of the network.
    2.2. The installation and operation of the climatological and precipitation stations will be according to the norms of the World Meteorological Organization.
    2.3. The exchange of short and long-term meteorological forecasts by means of the Regional Meteorological Centers of Brasilia and the National Meteorological Service of Buenos Aires, projected in the world meteorological vigilance plan.
    ...
3. To increase efforts towards the development of stations of common interest with equipment to make possible and/or more simple the collection and the exchange of hydrometeorological information.
4. To establish that the States limited by contiguous rivers (shared stretches) reach agreements to allow anyone of them to perform discharge measurements, over time and location, for a better definition of the fluvial regime in the border stretch, using non permanent and removable equipment and installations.
5. That in the contiguous rivers (shared stretches), when it is convenient for the riparian countries, to install and operate at selected section or

sections, border hydrometric gauges, to be observed by the respective State, periodically exchanging results of the observations.

6. That in all the border sections of successive rivers (non-shared stretches) to install water-stage recorders, for the continuous graphical recording of river height fluctuations in the section, and where the operating State meets the neighbouring State, over a period of time to be agreed, providing copies of the observation records.

7. To establish norms, by the joint action of the experts of the Basin countries, to set field observation methods, observation recording, statistical analysis techniques, regime characterization criteria.

   . . .

9. That all the processed data be divulged systematically by means of publications exchanged between them.

10. That the data to be processed, whether simple observations, readings or instrument graphical records, be exchanged or supplied according to the decision of interested parties.

11. As far as possible, to exchange gradually the cartographic and hydrographic results of their measurements in the Plata Basin, in order to facilitate a description of the dynamic system.

Considering

that water pollution control is an integral part of a water resource conservation and utilization policy and that any development programme of the water resource should always anticipate effective control of that pollution;

that it is important to conserve as much as possible the natural quality of surface and underground waters, protecting them from new contamination and to try to reduce the present contamination to guarantee their multiple purpose use, including potable and industrial water supply, irrigation and animal consumption, the conservation and development of flora and fauna and recreational aspects;

that in the function of the several uses anticipated for the Basin water resource, the concept of water pollution, which is in flux and therefore variable in space and time, should be well defined;

that a greater exchange of experiences of technical, administrative and legal character in the field of water pollution control is highly advisable;

Suggests to the Coordinating Intergovernmental Committee the following:

1. Legal and administrative aspects:

   1.1. A recommendation to the States to exchange information relating to administrative structures of the agencies in charge of the pollution control problem in each country, indicating areas and adopted methodology of work, as well as information concerning legislation in force.

1.2. To entrust to a group of experts a comparative study of the existing legislation and administrative structures (national, state, provincial, and municipal), that deal with water pollution in the different countries of the Plata Basin and, if possible, the presentation of a common text that can be accepted unilaterally by the respective States.

1.3. To entrust to the group of experts mentioned in the preceding paragraph an examination of and subsequent report on those practical procedures and international legal formulæ intended for preventing the pollution of surface and underground waters within the Plata Basin.

2. Technical aspects:
...

2.2. A recommendation to the states to exchange technical information, methods and technological procedures used by them in the field of water pollution control.

2.3. A recommendation to the States to initiate the establishment and development of research centres, in charge of studying and improving treatment methods for domestic and industrial sewage, as well as of training personnel and promoting the interchange of experts and information in this field.

2.4. To entrust to a group of experts the examination and definition of the following recommendations of a technical nature with the purpose of implementation:

2.4.1. Setting of uniform sampling and analysis methods to be used in the programmes for pollution control of the Basin waters.

2.4.2. Setting of a minimum number of significant parameters, to be exchanged, to allow a global appreciation of the problem.

2.4.3. Selection of research laboratories to serve as technological support for the development of those programmes.

2.4.4. Establishment of a number of sampling stations adequately distributed, that allow the follow up of programmes of pollution control of the basin waters, especially in the bordering areas of each country.

3. Programmes of pollution control:

3.1. Recommending the States to organize and execute, within their capabilities and according to the institutional and administrative structures, a programme of effective water pollution control in their territories.

3.2. Recommending that the States promote the development of educational programmes in order that the population understands and appreciates the importance of the conservation of the Basin water resources and the advantages resulting from adequate pollution control.

    3.3. Recommending that the States initiate the necessary facilities for the construction of installations for the treatment of domestic and industrial sewage, as well as the adoption of measures intended for divulging technological procedures looking for the reduction of the costs of those installations.

    3.4. Recommending that the States promote and encourage research and programmes whose aim is the repopulation of the ichthyologic fauna affected by the pollution of the Basin waters, as well as, in zones where it is required, the re-forestation intended for diminishing or eliminating the risks of the physical contamination of those waters.

Considering:
that the river's economic hydroelectric exploitation contributes to promote the development of the countries.

Suggests to the Intergovernmental Coordinating Committee that advises the Basin States:

1. Promotion of the completion of bilateral joint studies of the bordering stretches, to identify the best technical-economical solutions with a view to the hydroelectric development of the water resource in those stretches, taking into account other present and future uses of that resource;
2. That the construction and operation of hydroelectric works in non-shared reaches should not cause appreciable harm to other States of the Basin.

The third meeting of the group of experts was carried out in two parts. The first one took place between 29 June and 2 July 1971, and the second between 27 and 30 October 1971. The conclusions of that meeting are particularly significant:

The Group of Experts, considering:
that each State holds sovereignty over the stretches of international rivers of successive courses that flow through its territory and can adopt there the measures that it considers convenient to its national interest;

that that legitimate holding of its powers cannot imply detriment to the territorial integrity of other States that participate in the same hydrographic basin, by usage that causes them appreciable harm;

that has also been recognized by the Declaration of Asunción on the utilization of international rivers, when stating in its second paragraph: "In the international rivers of successive course, not having shared sovereignty each State can make use of the waters according to need provided it does not cause appreciable harm to any other State of the Basin";

that the effective respect to the terms of the Declaration of Asunción cannot refer to the construction stage of hydroelectric works but only to the consequences deriving from their operation;

that Article V of the Treaty of the Plata Basin states: "The collective action among the Contracting Parties should develop the implementation of those projects and undertakings they decide to carry out in their respective territories, having respect to international law and according to good practice among neighbouring and friendly nations";

that in numerous inter-American instruments it is affirmed and reiterated that the principle of good neighbourhood is in force, and its strengthening must govern all the efforts for the integral development of the Plata Basin resources;

that the effective application of this fundamental principle assumes, in relation to the State that makes use of the waters of a shared river, in the stretch under its jurisdiction, the duty of preventing and avoiding reasonably all appreciable harm that may be caused to other States of the same Basin and, in relation to them, the duty of bearing the minor inconveniences that may be derived from the operation of those developments;

that respect to all these principles should be completed with that of adequate publicity; and where that publicity is concerned, a practice such as the one adopted on the occasion of the filling of Jupiá dam (Brazil) is considered fundamentally satisfactory in view of the results obtained;

Thus it suggests to the CIC:
To recommend to the States that carry out hydroelectric exploitation of waters in international rivers of successive course in the stretches under its jurisdiction that, in respect to the operation and filling programmes of the reservoirs of works of that nature, to follow practices analogous to those performed in the case of the Jupiá dam (Brazil) for publicity on technical data relative to those programmes and adjust them to the rules and principles announced above.

The approval by the CIC of the preceding recommendation was accompanied by statements by the delegates from Brazil and Argentina, intending to clarify what "the practice of Jupiá" indicated in that recommendation consisted of. Since it is demonstrative of the differences in opinion existing at that time, it is interesting to analyse the details. In the meeting of the CIC of 22 March 1972 (Minute No. 152), the Representative from Brazil said:

... Actually what has happened is that:
1. in the case of Jupiá, knowing the situation of the navigation at Posadas, the Brazilian Government, for decision taken unilaterally, lengthened the chronogram for the filling of the dam and could ensure retention of navigation in Posadas, by means of a minimum discharge in the river, during the filling stage, of almost 3,050 m^3/s.

2. the Brazilian Government, at that stage, had information about the issue always published in important newspapers of Buenos Aires and São Paulo, and thus it seems that if the CIC condescends to approve the Resolution that is presented with a suggestion in that sense, so it should occur in the future.

On the other hand, the Argentinian Representative expressed (Minute No. 155):

... my Delegation wishes to complement the declaration of the Ambassador of Brazil with the following, in order to give an accurate idea to the Representatives of what actually occurred on the occasion of the filling of Jupiá dam. The declaration says:

1. Invited by Eletrobrás, and in an atmosphere of extreme goodwill, the President of the National Commission of the Plata Basin of Argentina, Captain Oscar Luis Lava, and its Technical Director, Engineer Julio Fossati, visited during the period 19–28 August 1968, the following Brazilian hydroelectric powerplants: Furnas, Estreito, Jupiá, Ilha Solteira and Capivari-Cachoeira.

2. The visit to Jupiá was made on 28 August 1968. In the preceding visits, with respect to the now completed laboratory and construction site, the Brazilian technicians had already briefed their Argentinian colleagues on the method to be adopted for filling the reservoir.

3. One and a half months later, on 3 October 1968, Captain Oscar Luis Lava and engineers Julio Fossati and Alberto Viladrich visited Eletrobrás in Rio do Janeiro and informed directors and experts of the Brazilian company of their concerns about the filling of Jupiá, during the critical drought period the region was experiencing. Based on data relative to the discharge measured at Posadas city for three typical years, they feared that an interruption to navigation could occur there for ships with a draught between 4 and 5 ft if the discharge of the Paraná river, at Posadas, was less than 7,000 m^3/s.

4. After a collaborative study, between Brazil and the invited Argentinian engineers, which considered the issue in all its details, in a conciliatory and cooperative atmosphere the Brazilian energy authorities determined the following measures: (a) The CESP would allow at Jupiá, during the filling of the reservoir, a discharge of not less than 3,050 m^3/s, which would ensure permanent navigational passage of between 4 and 5 ft of draught at Posadas; (b) the filling of the reservoir would be given extra time over that anticipated, to take into account the conditions of inflow and outflow discharges; (c) to facilitate follow up and study of observations made, and besides informing the local riparian populations, there would be daily publication of the observed discharges at Jupiá and Posadas in newspapers of São Paulo and Buenos Aires with large circulations.

5. When adopting these measures, the Brazilian authorities took into account the concern of the Argentinian technicians and the studies already mentioned, according to which it had been proved that, in October and November of the driest years of the period 1931–1959 (selected as the representative driest period), the navigation for 5 ft of draught had been interrupted for about 40 per cent of the time in the dry years of the same period. They were also based on the fact that in those months generally dry, very high discharges in Posadas, with relatively low ones in Jupiá, can occur.

The filling of Jupiá started on 25 November 1968 and the reservoir reached its normal operation elevation (280 m) on 22 January 1969. During this period the CESP always allowed discharges greater than 3,050 m³/s, with the exception of the last two days, 21 and 22 January, during which discharges were 2,680 and 2,770 m³/s.

Subsequently, the CIC was permitted to continue convening the same groups of experts for the projects presented at the second meeting of chancellors at Santa Cruz de la Sierra, mentioned before. Thus, at their fifth meeting, held at Punta del Este, in December 1972, project A-7, Ichthyological Resources was initiated, and subsequently, among others, Project A-4, Navigation.

Bilateral and trilateral agreements (1973–1983): working groups in basic areas

It was in 1973 that a push toward dispute solving and the realization of joint projects was begun, bi- and trilaterally. On 26 April 1973, Brazil and Paraguay subscribed to the treaty where the Itaipú Binational was created, and on 19 November of the same year, Argentina and Uruguay signed the Treaty of the Plata River, creating two commissions, in charge respectively of the river administration and of its maritime front. On 3 December the Yacyretá Binational Entity was originated, by agreement between Argentina and Paraguay (Barberis, 1988).

Subsequently, in February 1975, Argentina and Uruguay constituted the Administrative Commission of the Uruguay River. In 1980, Brazil and Argentina agreed upon the use of their shared stretch of the Uruguay River and decided to build the Garabí dam as a joint project.

These international treaties caused a substantial change in the institutional system of the Plata basin, linking the organization established by the Brasília Treaty with the system of binational commissions and entities mentioned, to which should be added those

133

established before 1973: the Joint Technical Commission of Salto Grande, created by Uruguay and Argentina to carry out their joint project of 1946, and the Argentinian-Paraguayan Joint Commission of the Paraná River – in charge of the administration of their shared stretch and of the Corpus Christi project – in 1971.

A particularly significant fact during this period was the signing, on 19 October 1979 – by the Governments of Argentina, Brazil, and Paraguay – of the Tripartite Agreement on Corpus and Itaipú, with the purpose of establishing rules in order to harmonize the Brazilian–Paraguayan development of Itaipú with the Argentinian–Paraguayan of Corpus, both on the Paraná river.

This agreement was the result of two meetings of a technical nature, held at Asunción on 22 and 23 September, and 17 and 18 November 1977; of two meetings of a diplomatic nature, on 14 and 15 March, and 27 and 28 April 1978, and of a negotiation process, on firm technical grounds, prepared by the delegations of the three countries involved. It can be said that the agreement has historical value since it put an end to the controversy concerned with the Paraná river energy utilization.

Parallel to this intense bilateral and trilateral activity, work in the ambit of the CIC was continuing. At the seventh chancellors' meeting held at Cochabamba in May 1975, it was decided to proceed to grouping CIC action into basic working areas, corresponding to the "Water Resources and other Natural Resources No. 2," comprising the current groups of experts for: "A-4. Navigation"; "A-7 Ichthyologic Resources" and "Water Resource."

The Working Group of Basic Area 2 (WGBA2) met from 2 to 11 November 1976, proposing a number of recommendations, among which should be noted the preparation of a glossary of terms about water-resources pollution. Subsequently, in a meeting held in October 1977, members established the parameters which were to be followed for the assessment of water quality, the techniques to be used and the chronogram of activities.

Simultaneously the CIC convened the Subgroup of Ichthyology Specialists of the WGBA2 at a meeting in October 1978, in order to discuss the "development of a project creating a Center of Hydro-biological Research for the Plata Basin." The subgroup prepared the corresponding terms of reference and recommended that the CIC convened a meeting of specialists during the first part of 1979, in order to draw up an outline for the project, together with the Group of Specialists in Hydrobiological Resources and Fishing Legislation.

In July 1980 the WGBA2 met and constituted a subgroup of hydro-biology, which analysed the items that would comprise a programme of hydrobiological research and suggested to the CIC that a programme should be prepared with a common methodology with definition of the areas to be given priority: (1) ichthyologic resources – inventory of ichthyofauna, fishing biology, water culture; (2) continental aquatic ecology; and (3) water quality.

In July 1982 the Subgroup on Water Quality of the WGBA2 met, and recommended that the CIC should ask the countries concerned to increase their efforts to comply with Resolution 140(XI) of the eleventh meeting of chancellors (Buenos Aires, December 1980). That resolution was devoted to the following aspects: (a) to determine the water quality parameters to be incorporated in the control scheme; (b) to adopt the appropriate methodologies for analysis or determination; (c) to select in each country one or more reference laboratories.

With respect to activities in hydrometeorology, in April 1978, the CIC decided to ask countries to declare the national centres appointed for the reception and transmission of hydrometeorological information. In October 1979 the corresponding group of specialists met, in order to prepare a project for the organization of a Hydro-meteorological Documentation and Data Centre for the Plata basin. At that meeting, the specialists decided to recommend to the CIC that it should request OAS to update the *Inventory of Hydrological and Meteorological Data of the Plata Basin*, published in 1969.

In July 1981, the WGBA2 dealt with the subjects with which the CIC had been entrusted by the chancellors at their eleventh meeting, held in Buenos Aires in December 1980, which comprised the coordinated studies to be carried out with the aim of understanding the behaviour of the basin's water system, in relation to the Paraguay river floods and the actions for mitigating their effects.

Following a proposal by the Uruguayan delegation, it was decided to extend the scope to the remaining rivers of the basin, and it was recommended to the CIC that it should alert the governments to the necessity for the appointment of national centres for the reception and transmission of hydrometeorological information, having agreed the procedures and programmes for the exchange of such information, and also information from the countries about systems in use and proposed for flood warning and organization of civil defence that should be clarified with other member countries.

In July 1982, the WGBA2 called the CIC to convene a meeting of specialists in order to clarify the need for member countries to show respect for the establishment of an adequate hydrometeorological network, taking as fundamental guidelines for the updating of a hydrological and climatological data inventory to be made by the OAS and to study the compatibility of national data banks.

Revision of the Basin Institutional System (1984–1996): Technical Counterparts

At their fifteenth meeting held in December 1984, the Ministers of Foreign Affairs subscribed to the "Declaration of Punta del Este," by which they convened an extraordinary meeting of Undersecretaries or Special Representatives of the Ministers of Foreign Affairs, with the aim of analysing and evaluating the institutional political state of the system of the Plata basin. That extraordinary meeting had to identify: (a) the achievements and accomplishments reached; (b) the decisions adopted which had not been thoroughly executed; (c) the difficulties faced by the system from a substantive and operative point of view, given their origin and nature. Additionally, it was to examine "whether the present structure and organization of the system bodies are the most appropriate for the attainment of objectives and purposes of the Treaty of the Plata Basin."

The CIC constituted an ad hoc group with the aim of complying with the preceding subjects (a), (b) and (c), which held sessions between July and October 1985 and drew up a diagnostic report for the performance of the System, identifying orienting criteria which would contribute to its reactivation, and including a survey of achievement and accomplishment reached within the spirit of the treaty and the stage reached in the Resolutions of the Ministers' meetings.

The Extraordinary Meeting of Undersecretaries or Special Representatives of the Ministers of Foreign Affairs of 18 and 19 November 1985, took the guidelines of that report and agreed with the necessity of reactivating the System on the basis of a pragmatic approach that gave it greater operating agility, allowing the concentration of efforts in four important subjects: "Water Resources and Other Natural Resources," "Navigation," "Fluvial and Terrestrial Transport" and "Border Cooperation."

The Resolution No. 2 (II-E) of the second extraordinary meeting of Ministers, held in Buenos Aires in April 1986, established:

Considering:

That the meeting of Chancellors has reaffirmed the objectives and purposes of the Treaty of the Plata Basin as instrument of integration, cooperation and development of their people and agreed to analyse and evaluate the political institutional state of the System;

That the report 1/85 approved by the CIC defined basic criteria for the future performance of the System

- of the pragmatism, with the aim of facilitating cooperation among the Member Countries by means of the utilization of more effective operative mechanisms;
- of an intensified technical cooperation based on direct understandings among the national organizations competent on subjects previously selected;
- of the concentration of joint efforts in the subjects of priority interest for the better fulfilment of the objectives of the Treaty;
- of the operative flexibility, by means of the concentration of specific or partial bilateral or multilateral agreements, for the participating Member Countries.

That from examination of the present structure and institutional organization of the System it is convenient for the CIC Secretariat to perform technical tasks of support, promotion and coordination in relation to the activities of reciprocal cooperation that the Member Countries develop.

Resolves:

1. To implement the basic criteria included in the Report 1/85, approved by the CIC, whose pragmatic approach will allow its instrumentation and improvement according to developments.
2. To recommend that the specific or partial, bilateral or multilateral agreements reached according to the Article of the Treaty, be open, as necessary, to fulfilment by other Member Countries, thus giving value to the dynamic potential that this represents for actions towards the accomplishment of the general objectives of the Basin.
3. To concentrate the joint cooperation efforts of the Member Countries in those priority subjects that allow a greater effectiveness of the associative tasks developed in the ambit of the Treaty of the Plata Basin. Priority will be given to the subjects identified in the Report 1/85 (Water Resources and other Natural Resources, Navigation, Fluvial and Terrestrial Transport and Border Cooperation). This operational priority will not have excluding character of the other subjects specified in the Unique Paragraph of Article 1 of the Treaty ...
4. To generate technical cooperation by means of direct understanding among the national organisms competent in specific subjects. For that purpose the Governments will make known to the CIC the responsible counterparts in each case.

Accordingly it was decided, at the sixteenth meeting of Ministers, held at the same time as the second extraordinary meeting, to entrust to countries nominated to the CIC Secretariat, the national organizations which would act as "technical counterparts" in selected areas and, putting into practice particular projects, decided to establish a Hydrological Warning System. To that end the CIC would entrust its Secretariat with convening a special meeting of the technical associates designated by the contracting parties.

It also decided to ask governments to appoint and communicate to the CIC the national organizations that would act as technical counterparts, with the aim of proceeding to the evaluation of water quality in the basin's international rivers.

The deliberations of the report of the second extraordinary meeting led also to the adoption of the "Declaration of Buenos Aires" on 4 April 1986, by which it was decided to drew up a "Programme of Concrete Action" consisting of activities in progress and identifying a set of specific projects, regarding the general priorities indicated in Resolution No. 2 (II-E) in a pragmatic manner.

The programme was approved in December 1987, during the seventeenth meeting of Ministers, held in Santa Cruz de la Sierra, and was put into practice the following year. It comprises the following projects related to water resources:

I-1   Hydrological Warning System
II-1 Water Quality Control of the Basin Waters
III-1 Regional Programme of Soil Conservation
III-2 Regional Documentation Centre on Soil Resource and other Natural Resources
IV-1 Signalling Navigable Ways
IV-2 Improving Navigable Ways
IV-3 Network of Coastal Stations and Information about Navigation
IV-4 Facilities for Navigation.

For each of them, global objectives, short and medium-term targets, present conditions, and a working plan were established.

The Technical Counterparts on Hydrological Warning met five times, in October 1986, April 1987, June 1988, April 1990 and April 1993. Between 20 and 22 October 1986, they had met for the first time in order to analyse the operational start of a hydrological warning system.[3]

At that point the institutions and organizations of each country would be responsible for the exchange of data with a view to the implementation of the establishment of operative centres for a warning network. The stations that would constitute the network initially – in the Bermejo, Pilcomayo, Paraguay, Paraná, Iguazú, and Uruguay river basins – were agreed, on both the parameters to be exchanged and the frequency of transmission in normal and critical or particular situations. At the same time, interest was expressed in having available forecasts of heights and/or discharges at different sites in the basin and it was decided to exchange the results of forecast models already in existence. The necessity for resorting to FONPLATA or some other financing source, with the objective of enlarging or improving the network infrastructure or the information transmission system among responsible organizations was also taken into account.

Between 20 and 22 April 1987 the second meeting was held, with the aim of evaluating the System's performance and adopting the necessary adjustments for its improvement.[4] A report was drawn up about the problems which had occurred since the preceding meeting, the project stages reached for the installation of new stations, eventual modifications to the network and the possibility of increasing the number of stations. It was agreed monographs and historical information about the stations would be produced and procedures to be established in the event of extraordinary circumstances. It was also recommended to the CIC that there should be priority interest in the financing of projects for enhancing equipment and installations and promoting the interchange of technical visits between countries.

The third meeting was held on 27 and 28 June 1988.[5] After exchanging information about the state of the Warning Network and evaluating its performance since the preceding meeting, the study of mechanisms for requesting financing for the System through FONPLATA was instigated. Modifications in the Network were also discussed with the frequency of transmission of information on floods, the sending of forecasts and the exchange of historical information, concluding with the advances achieved in the correct orientation of the Programme of Concrete Actions.

The fourth meeting was held between 25 and 27 April 1990. The main subjects of the Agenda[6] were the exchange of agreed hydrological information; eventual modifications in the Network; frequency of transmission of information; supply of forecasting; transference of know-how; analysis of specific projects and their financing, ending with the Annual Report on the progress and evaluation of Project I.1 of the Programme of Concrete Actions.

The Technical Counterparts for Water Quality met in October 1986, April 1987, June – July 1988, April 1990, August 1991 and March 1992 (both extraordinary meetings) and April 1993. Their first meeting was on 23 and 24 October 1986, commissioned with recommending measures to be adopted with the aim of assessing the water quality of the international rivers of the basin, as well as those measures for the prevention and fight against their pollution.[7]

After the presentation of antecedents about the works carried out in each of the countries and on institutional and legal aspects, it was decided that each country would make an evaluation of the available information on quality criteria for different uses (supplies for human consumption; previous conventional treatment; recreation activities with direct contact; agriculture and cattle breeding activities and conservation and development of aquatic life), taking into account the parameters approved formerly by the Ministers' Resolutions. That evaluation will decide the definition of the basic aspects of the design of a quality evaluation system, comprising parameters to be measured, sampling frequency, stations for taking samples and operative mechanics for the collection and processing of information.

At the second meeting, held on 23 and 24 April 1987, it was decided to adopt a Minimum Evaluation Network, in which samples would be taken quarterly at the first stage, following a tentative operating mechanism established by guidelines drawn up by delegates.[8] It was also agreed to recommend the declaration on priority interest for the financing of projects for the establishment of selected sampling stations and strengthening the interchange of technical assistance and cooperation.

The third meeting took place between June 29 and July 1 1988.[9] After reporting the monitoring tasks at different stretches of the basin rivers, observations were made in relation to the operative situation of the Minimum Network adopted at the preceding meeting; it was decided to keep the quarterly sampling frequency if possible; it was also suggested that new parameters to the list in force up to that moment should be incorporated; some proposals were presented in relation to the adoption of joint operations, and talks began on information exchanges in emergency situations.

The fourth meeting was held between 18 and 20 April 1990, and the constitution of a Working Group in order to prepare a "Methodological Guide for the Operation and Evaluation of the Plata Basin Water Quality Network" being an important item on the agenda with the adoption of common guidelines and recommendations for the drawing up of contingency plans in each country and a global plan for the basin itself.[10]

From 5 to 8 June 1990, the Working Group met with the aim of discussing and approving a very detailed proposal for the methodological guide, as previously decided.[11] A common text was agreed, to be considered by the Brazilian representative, who was unable to be present at the meeting.

The guide was approved at an extraordinary meeting of the technical counterparts held on August 12 and 13 1991, and its recommendation considered at the following Ministers' meeting.[12] After approval, the Ministers decided to present it as a contribution to the United Nations Conference on Environment and Development, held in Rio do Janeiro in 1992.

On March 19 and 20 1992, delegations designated by the member countries met at another extraordinary meeting, having as its specific subject the problem of cholera, which was affecting several zones of the basin. The meeting's agenda comprised the following items: vigilance programmes adopted; methods used for the detection of the *vibrio choleræ* in the water environment; identification of the discharge of sewage in the basin rivers (location, mean daily discharge, type of treatments applied); qualified laboratories network in each country; warning system utilization by organizations appointed for the notification of pollution incidents.

In December 1992, given the necessity of "adapting the CIC and the Technical Counterparts to the new reality presented by the sub-regional integration process," the Ministers decided to approve a new statute for the CIC giving it, among others, the following attributions:

– To consider, approve and implement projects, study and research plans referred to in the only paragraph of Article 1 of the Treaty of the Plata Basin and specially those related with the Programme of Concrete Actions, determining its priorities.
– To update and reform the Programme of Concrete Actions.
– To convene, with specific mandate, meetings of Technical Counterparts or working groups ...

For the first time after the structural modification of the CIC, from 28 to 30 April 1993, a Joint Meeting of Technical Counterparts on Water Quality and Hydrological Warning was held, with participating representatives of international organizations of technical and financial cooperation with competence in those areas (OAS, UNESCO, PHO/WHO, WMO, FONPLATA, IICA).[13] Representatives of international and regional cooperation agencies reported on programmes in progress in their respective organizations, having been asked about the feasibility of participation and financial support for preliminary projects resulting from the deliberations.

Separated into two working subgroups, the technical counterparts prepared two preliminary projects devoted to the immediate creation of national and regional response capacities. One referred to water quality and pollution control and the other to hydrological warning. By taking these preliminary projects as a basis, FONPLATA and IDB were approached concerning the necessary financing for their implementation. In spite of the fact that the April 1993 meeting coincided with a new stage in CIC development, and delegates expressed their satisfaction with the expectations generated, the technical counterparts have not as yet met again (December 1996). Bilateral and multilateral organizations, have, however, been very active.

A repetition of activity pattern of the 1970s decade, commented upon above, can be seen in the creation of the Administrative Binational Commission of the Lower Basin of the Pilcomayo River, by Argentina and Paraguay, in September 1993; of the Trinational Commission for the Development of the Pilcomayo River Basin, by Argentina, Bolivia, and Paraguay, in February 1995, and the Binational Commission for the Development of the Upper Basin of the Bermejo River and the Grande River of Tarija, by Argentina and Bolivia, in June 1995.

For its part, in the navigational field, the Intergovernmental Committee for the Waterway Paraguay-Paraná (Cáceres Port–Nueva Palmira Port) (ICW) has been active, with the participation of the five member countries, riparians of those rivers. In this case, it is interesting to notice that the Waterway Programme was incorporated into the System of the Treaty of the Plata Basin in October 1991, although keeping the structure of the ICW, whose statute was approved by the Ministers in December 1992.

Conclusions

This description has shown the praiseworthy efforts of experts in the different disciplines relating to water resources responding to the challenge of a basin with the complexity of the Plata one, trying to solve fundamental issues in the search of sound knowledge about the behaviour of its rivers, in normal and extraordinary situations, the most adequate way to utilize them, and the concern for preserving the basin's quality. This effort has been rewarded with very positive results, such as the uninterrupted performance during more than a decade of the Hydrological Warning System, always kept active through a daily response in addition to the occurrence of particular phenomena such as the 1992 floods.

It is also worth noting, among other things, the tasks carried out jointly to prepare the *Methodological Guide for the Operation and Evaluation of the Plata Basin Water Quality Network.*

However, it must be admitted that many difficulties have come up, and these difficulties have prevented the implementation of many very interesting proposals and projects, with some partial completions.

Certain limitations arise from the way the institutional system of the basin is organized. As Barberis (1988) notes, the system con-stituted on three levels (Conference of Ministers of Foreign Affairs, Coordinating Intergovernmental Committee, Groups of Experts, or Technical Counterparts) has not worked satisfactorily. One of its fundamental flaws is due to the lack of a permanent technical orga-nization. The CIC is a body composed of diplomats, whose training does not allow them to consider deeply the multiple subjects sub-mitted to their consideration and decision. This circumstance has been the cause, which soon became apparent, about the need to have a body which can provide scientific and technical knowledge for the Committee in order to adopt a political decision suitable to each case.

The former working groups and the present technical counterparts have not been able to make up for that failing. It frequently occurs that groups are not integrated by experts only, but the diplomats participate as well. This often gives discussion on each subject the characteristics of a negotiation, thus moving away from its scientific approach.

In addition, delegations often present a considerable number of documents and proposals, and this hinders the experts from carefully reading and analysing all the matter during the short time of the meetings. Since it is almost impossible to examine all the material presented, either decisions are postponed or they are negotiated, approving some documents from each country, thus preserving a certain political balance. Another flaw in the system lies in the fact that the resolutions of the meetings of Ministers and of the CIC, addresses to the member states have, in general, the character of mere recommendations and, consequently, they lack a legal obliga-tory force (Barberis 1988). To that it should be added that when the resolutions concern subjects foreign to the usual competence of the Ministers of Foreign Affairs, the technical bodies of each government generally do not pay much heed to their recommendations.

Another flaw has been the lack of specific funds for the financing of the programmed activities. The results obtained are only possible because of the economic contributions of the organizations where the intervening experts come from and, generally, they are not included in a budget specially assigned to those activities. Nor could the promised finance from international agencies be obtained owing to the absence of concrete projects with an adequate level of development. This is also a consequence of the fact that the experts have only been able to dedicate a small amount of time to the tasks agreed in the respective meetings. The development of the Plata basin established by the Treaty of Brazil has lost its original vigour and the works presently carried out in the basin are mainly the fruit of the activities of the bilateral or multilateral commissions or entities.

In order to revitalize the Basin System, in carrying out the original objectives conceived by the Ministers at their first meetings, it is necessary to look for solutions to the obstacles stated above but that search must take into account the new realities in which the member countries are living. In the first place, there is the great integration project, the Southern Common Market (Mercosur), created by agreement of four of the five basin countries – Argentina, Brazil, Paraguay, and Uruguay – with the fifth, Bolivia, soon to be incorporated.

This project, which cannot yet be clearly visualized, has advanced principally in the commercial field and has not faced up to other problems, such as the use and preservation of the basin rivers, as an integral concept, although it is analysing sectoral issues closely related, such as the energy integration of the member countries.

Another question to be considered is the transference to the private sector of activities which for many years have been in the charge of the states. So far, the hydraulic works built in the Plata basin have been built and have been generally operated directly by the states or by state-owned companies or by binational entities constituted by the states. Private companies have only participated as contractors.

The privatization process carried out in Argentina recently and that will possibly be repeated to a greater or lesser degree in the other countries of the basin, has meant that the Argentinian national hydroelectric plants are being operated by private consortia and that there are proposals for privatization – whose characteristics are presently in discussion – of the binational developments Yacyretá and Salto Grande.

If one adds the studies carried out by the Argentinian-Paraguayan Joint Commission of the Paraná river about the possibility of offering the construction and subsequent operation of the Corpus Christi development for privatization and similarly, the recent call for bidding by the Binational Commission for the Development of the Upper Basin of the Bermejo River and the Grande River of Tarija, for their developments Las Pavas, Arrazayal, and Cambarí, the existence of a new reality, certainly unimagined when the basin institutions were created, was to be envisaged.

In the field of navigation, as well, active private participation can be anticipated, having already taken a first step by giving the concession for the dredging, maintenance, and buoyage of the Paraná river between Santa Fé and Buenos Aires to a private consortium.

In the water supply and sanitation sector the tendency is also towards different forms of private participation. This process of privatization implies the reduction or disappearance of state entities, among which can be found, for example, those responsible for the operation of the hydrological networks. On the other hand, it means the incorporation of new actors, the concessionaire companies and the respective regulating and control entities. In addition, the new world approaches in relation to water resources should be added to this new regional reality. It is noticeable that a reappraisal of water has become an issue, as old problems arise with new aggravating conditions, such as those related to scarcity in many areas of the planet. That reappraisal was made evident at international meetings, particularly the International Conference on Water and the Environment (Dublin, January 1992), the United Nations Conference on Environment and Development (Rio de Janeiro, June 1992) and in the programmes of international credit agencies.

If the advantages offered by technological advances as regards communications and tools for collection, processing, and presentation of information are also taken into account, there is a favourable outlook for facing the revision and renewal that the Plata basin system requires. So this is a propitious moment to analyse critically the existing institutional mechanisms and from that analysis, modifying them when necessary, to improve their effectiveness.

The challenge to be faced is very large. All the positive actions that have enabled countries making up the basin to speak the same "hydro" language should be supported. But it is necessary to go further, to be more ambitious, if recovering the concept of the basin as an unit and facing future development by working together in an integrated way is the intention (Pochat, 1993).

Mechanisms to facilitate that joint work, that help to complement efforts, link disperse experiences in many fields, and plans with a comprehensive vision of the basin's potentiality must be envisioned, with the new regional and global realities, and the diversity of demands to be satisfied. And it is necessary to advance beyond diplomatic contacts, certainly very important, since they open many doors, but not enough if they are not converted into concrete activity.

The technical experts have always had the opportunity to respond with greater fluidity, because they have been able to analyse the problems, playing a perhaps less compromising role, and besides, they have had the chance and obligation to look at each problem in detail, in search of a practical solution. That comparative advantage should be profitable, when they sit at the same table to plan the common future. In that way the technicians of the five countries of the Plata basin will be paying back, at least partially, the privilege of being engaged with this exceptional and vital task.

Notes

1. Minutes of the meetings of the CIC Nos. 96–99, 19 March, 2, 7, and 24 April 1970.
2. Notes No. 8/851 (40a) of 20 April 1970, from the Brazilian representative and No. 320/970 of 21 April 1970, from the Argentinian representative.
3. Report of the Meeting of the Technical Counterparts of the Countries of the Plata Basin on a Hydrological Warning System. Resolution No. 195 (VXI), 22 October 1986.
4. Ibid., 22 April 1987.
5. Ibid., 28 June 1988.
6. Ibid., 27 April 1990.
7. Report of the Meeting of the Technical Counterparts on Water Quality. Resolution No. 196 (XVI), 24 October 1986.
8. Ibid., 24 Apil 1987.
9. Ibid., 1 July 1988.
10. Ibid., 20 April 1990.
11. Report of the Working Group for Drawing Up *The Methodological Guide for the Operation and Evaluation of the Water Quality Network*, 8 June 1990.
12. Ibid., 13 August 1991.
13. Report of the Technical Counterparts on Water Quality and Hydrological Warning, 30 April 1993.

References

Barberis, J.A. 1988. La Plata River Basin. Interregional Meeting on River and Lake Basin Development with Emphasis on the Africa Region. Addis Ababa.

Cano, G.J. 1979. *Recursos Hídricos Internacionales de la Argentina*. In: Víctor P. de Zavalía, ed. Buenos Aires.

Comisión Mixta Argentino-Paraguaya del Río Paraná (COMIP) 1992. *Aprovechamiento Energético del Río Paraná. Documentos y Tratados*. Buenos Aires.

del Castillo de Laborde, L.C. 1991. *El Tratado de la Cuenca del Plata, un Sistema en buscade su Definición*, 17 Curso de Derecho Internacional, Comité Jurídico Interamericano, OEA, Washington, D.C.

Pochat, V. 1993. *Gestión de Cuencas Internacionales*, 10 Simpósio Brasileiro de Recursos Hídricos y I Simpósio de Recursos Hídricos do Cone Sul, Gramado.

6

Environmental management issues in the Plata basin

Newton V. Cordeiro

Introduction

The Plata basin is a vast river system, second in size only to the Amazon river basin, whose natural resources are of critical importance to the economy and social welfare of South America. The basin is composed of three major rivers which drain some 3,100,000 km^2 – almost a fifth of the continent – to the Atlantic ocean. Almost half of the basin – 1.4 million km^2 (or 45 per cent) – is situated in Brazil. Another 30 per cent lies in Argentina, 13 per cent in Paraguay, 7 per cent in Bolivia and 5 per cent in Uruguay (figure 1).

This great basin contains a wealth of natural resources, particularly water and soil resources, mineral deposits, and large stands of forest. For each of the five countries, most of their developed agricultural and industrial regions lie within the basin, generating 80 per cent of their combined gross domestic product. There are important oil and gas resources, particularly in Argentina and Bolivia, and an extensive river and land transportation network, which links politically and economically all Plata basin countries.

In terms of water resources, the Plata basin includes some of the most important hydroelectric dams in Latin America (i.e. Itaipú,

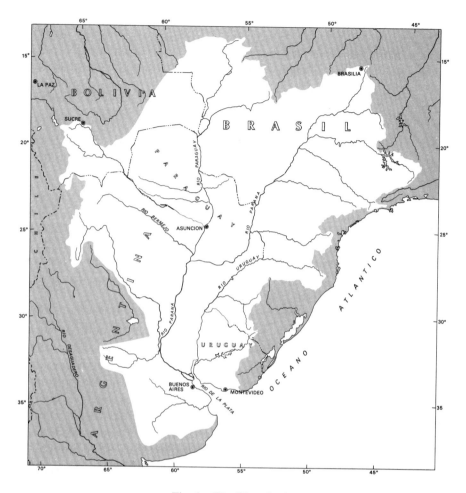

Fig. 1 **The Plata basin**

Yacyreta, and Salto Grande). These hydroelectric plants, constructed over the last 30 years, have raised the installed energy capacity to 42,000 MW and the volume of impounded water in reservoirs to about 350,000 hm^3, a third of which is usable for energy production and other water uses. The basin also contains one of the world's largest aquifers (Guarany), a 1.2 million km^2 hydro-stratigraphical system in the Paraná geological basin. This aquifer has reserves estimated at 50,000 km^3 of excellent quality water suitable for domestic and industrial use.

The Paraná, Paraguay, and Uruguay rivers are natural routes for the shipment of goods for export or for consumption along their courses. The Hidrovia Paraguay-Paraná, originating in Brazil and flowing downstream to Buenos Aires and Montevideo constitutes the most important international navigation route in South America's southern cone. This waterway is an essential component in the ongoing initiatives for economic integration and cooperation in the region and is also the centre of the discussion because of the potentially enormous negative environmental impacts that these initiatives could have throughout the basin.

Since 1969, most development and physical integration activities in the region have been bound by the Treaty of the Plata Basin, signed in Brasília in April of the same year. The treaty also instituted the Intergovernmental Coordinating Committee of the Plata Basin Countries (CIC) as a mechanism for promotion, coordination, and follow up of multinational activities and initiatives for the integrated development of the Plata basin.

However, population growth, large-scale economic projects, the economic crisis of the 1980s, and changes in development priorities and approaches have contributed to the degradation of the Plata basin's natural resource base, the consequences of which are seen today primarily as problems of erosion, sedimentation, and pollution.

The Plata basin: an environmental profile

As previously stated, during the last decades several factors have contributed to create serious environmental problems in the region. Among these, the most urgent are the erosion of productive land, the silting of waterways and reservoirs, soil and water pollution, floods, droughts, and loss of habitat for fish and wildlife.

Erosion

The impacts of erosion are evidenced primarily by the sedimentation of existing reservoirs and certain reaches of the river system. In some areas erosion has already reduced the river's transport capacity. There has also been loss of productivity of croplands and pastures, and the destruction of land suitable for reforestation. Increased sedimentation in the main rivers and secondary basins has increased the cost of dredging in the Plata river. Most severe cases of erosion in the region are caused by the expansion of the agricultural frontier.

Sedimentation

There are some areas of high sediment production in the Paraná and Paraguay basins. The most severe erosion occurs in the upper basin of the Bermejo River, a tributary of the Paraguay river, in the Andean region of Argentina and Bolivia. Another area of high erosion and sedimentation is the Upper Pilcomayo river basin in which 90 million metric tons of sediment are annually deposited on the wide flood plain of the Chaco producing an extensive inland delta. In the Upper Paraguay river basin in Brazil, most of the sediment from the headwaters, especially from the São Lourenço and Taquarí river basins, is deposited in the Pantanal. In the Paraná river basin the largest volumes of sediment are eroded from sedimentary rock, mainly sandstone, located upstream from the Itumbiara and Itaipú reservoirs.

Pollution

Water pollution is caused primarily by the great increase in crops and industrial production, and by population growth. A large part of fertilizers and pesticides used in farming is carried by runoff into watercourses. This toxic pollution is not only posing risks for populations that depend on the river's productivity for their livelihoods but threatens the biodiversity of the maritime front of the Plata river. Another significant source of pollution is untreated urban sewage.

Natural hazards

Damages from floods, landslides, and prolonged droughts, among other natural causes, are taking a rising toll in lives and property. The basin's rainfall, its agricultural and industrial development, the presence of numerous dams, and prospective transportation infrastructure projects in connection with the Paraguay-Paraná waterway enhance the vulnerability of the Plata basin to natural disasters.

Some of the main factors that have contributed during the last three decades to alter the Plata basin's environment and create ongoing problems are population growth, the expansion of the agricultural frontier, and the implementation of large-scale economic projects, particularly related to energy generation.

Population growth

The population grew from 61 million inhabitants in 1968 to about 116 million in 1994. Nearly 60 per cent of the total population of the five countries is living in the basin, with much of the population concentrated in small and intermediate cities. The urban population in the Plata basin increased from an average of 45 per cent at the beginning of the sixties to an estimated average of 77.5 per cent. In the state of São Paulo, Brazil, 93 per cent of the population is presently urban. Population growth and increase in social mobility resulting from migration from the countryside and the more impoverished areas to the urban areas have given rise to new metropolises, and several mid-sized cities. In the rapidly growing industrializing cities of the Plata basin many urban centres lack basic economic and social infrastructure. A concentration of poorly located industries, having no proper treatment facilities, increase organic and chemical pollution still further. Some of the greatest future environmental problems confronting the Plata basin countries will stem from this accelerated urban growth.

Expansion of agricultural frontier

In the Upper Paraná basin, because of the intensification of agricultural and industrial production, many areas formerly planted with coffee and food crops were converted to soybeans for export and sugarcane for the production of fuel alcohol. In the Paraguay river basin, in both the Brazilian and Paraguayan drainage areas, extensive tracts of forests were cleared for increased crops and pastures. In the 1960s, the Paraguayan Government, in response to mounting demand for employment and higher revenue, encouraged expansion of the agricultural frontier in the eastern region of the country, both in the Paraguay and in the Paraná river basins. As a result, the forests, which covered 45 per cent of eastern Paraguay at the end of the 1960s, shrank to about 35 per cent in the mid-1970s, to 25 per cent in the mid-1980s, and to only about 15 per cent at the beginning of the 1990s (Bozzano and Weik, 1992).

Energy

The hydroelectric power sector expanded, especially in the Paraná and Uruguay river basins. Here 15 hydropower plants of more than

1,000 MW capacity were or are being built, raising the installed capacity in the basin from 2,000 MW in 1966 to 42,400 MW in 1996. This availability of energy created a great expansion of industrial production and overall economic activity, especially in the north eastern part of the basin, in Brazil.

Besides these factors, during the last 30 years there were also significant changes in the socio-political and institutional context that drastically changed development priorities and approaches, contributing to the degradation of the Plata's natural resource base. During the 1970s, the political scene was characterized by the presence of military governments. The decade could be divided into two periods: the first part, influenced by river basin corporation models, was characterized by attempts to create secondary and tertiary development poles to expand urban and industrial development. The major projects targeted regional development, with particular emphasis in water resources and transportation development. Frustration with the results, the oil crisis, and new concerns about meeting the population's basic social needs, led to a second period in which regional strategies emphasized rural development.

In the 1980s, the economic crisis led the Plata countries to adopt new priorities and make profound changes in their mechanisms for coordinating social action. The external debt problem and the adjustment process caused the governments to turn their efforts toward increasing tax revenue, and giving priority to exports which resulted in improper management of their natural resources. The decade of the 1990s finds most countries consolidating their political and economic situation, and practically all of them engaging in the transformation of governmental structures and trying to reach a balance between the public and the private sector. In economic terms, growth resumed significantly, after the so-called lost decade. The three basic processes at the core of the trends being pursued in the Plata region are: changes in the role and organization of government leading to decentralization, adjustments in the countries' economic structures, and reintegration in the international economy. There is also a renewed and increased interest in regional economic cooperation and stability, reflected primarily and most obviously by the establishment of the Southern Common Market (Mercosur) in 1995 between Argentina, Brazil, Paraguay, and Uruguay; by an increased participation of the private sector in the economic recovery of the region; and by an accentuated interest in developing the Hidrovia Paraguay-Paraná, as an essential component in the initiatives for trade and cooperation and the renewed efforts toward stabilization and liberalization in the countries.

Critical areas for sustainable development activities

There are a number of distinct sub-regions or sub-basins of the Plata basin and their immediate areas of influence which face critical natural resources management problems. Most of these sub-regions or river basins are shared by two or more countries: the Upper Paraguay river basin, the Pilcomayo river basin, the Bermejo river basin, the Chaco region, and the Mirim Lagoon basin (figure 2). These sub-regions and sub-basins are described below.

The Upper Paraguay basin

The Upper Paraguay river basin (UPRB) is delimited as that portion of the Paraguay river basin above the confluence of the Apa river which forms the border between Brazil and Paraguay (figure 3). This basin drains an area of 496,000 km^2; 396,000 km^2 of which are located in the states of Mato Grosso and Mato Grosso do Sul, Brazil, and the remaining 100,000 km^2 are located in Bolivia and Paraguay. Within this area lies the Pantanal, the most extensive wetland ecosystem in the world. It comprises an area of approximately 140,000 km^2 – an area large enough to encompass Austria, Belgium, Hungary, and Portugal. The Pantanal has been identified as an area of special concern by the World Conservation Union (Dugan, 1990),[2] and proclaimed to be an area of National Heritage in the Federal Constitution of Brazil.

The Brazilian portion of the Upper Paraguay river basin (UPRB) encompasses two interdependent ecosystems: the upper sub-basin (or Rim), comprising about 256,000 km^2 at altitudes above 200 m, and the lower sub-basin (or Pantanal). The lower sub-basin receives its water from the upper sub-basin, and remains flooded for several months, each year.

The Pantanal occupies about 35 per cent of the total watershed area of the UPRB in Brazil. It is a very flat region. In several areas, flows reach numerous small lakes through abandoned meanders which have been cut off by the main channel. These lakes are generally covered by floating vegetation. As a result of these processes, the Pantanal behaves like a reservoir which retains a large portion of the total annual runoff. It has been estimated that the total sub-

PLATA BASIN

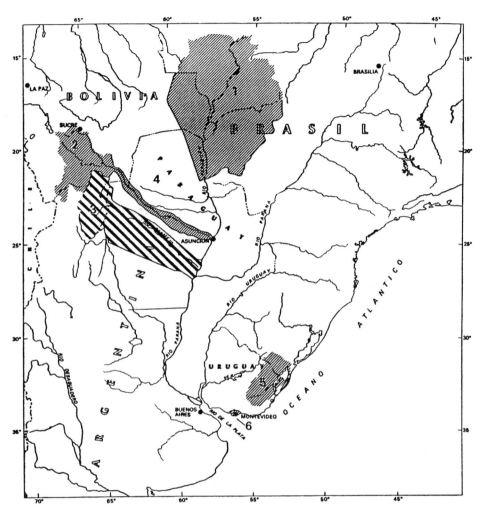

CRITICAL AREAS FOR SUSTAINABLE DEVELOPMENT ACTIVITIES

1	Upper Paraguay River Basin	4	Gran Chaco
2	Pilcomayo River Basin	5	Laguna Merin
3	Bermejo River Basin	6	Bay of Montevideo

Fig. 2   **The Plata basin: Critical areas**

Fig. 3 **Map of physiographic regions and main drainage system of the Upper Paraguay river basin (Project RADAMBRASIL, 1984)**

merged area can vary annually from 10,000 km² to 30,000 km². Maximum annual precipitation occurs close to the divide between the UPRB and the Amazon river basin. An extensive network of rainfall, river flow, and water quality gauging stations is present in the basin.

Due to its location at the centre of South America, the Pantanal is a bio-geographical meeting point of several endemic floral and faunal units. The confluence of elements of fauna and flora with origins in the Amazon, Chaco, Savanna, and Mata Atlantica regions contributes to the extensive biological diversity of the region. The Pantanal also displays different types of upland and lowland stationary forests, the "chaquenhas" as well as a diversified species of savanna known as "cerrados" – open areas of steppe-like grassland, and natural pastures. The Pantanal is habitat for a unique and extremely rich array of wildlife: more than 230 species of fishes, 80 species of mammals, 50 species of reptiles, and more than 650 classified species of aquatic birds.

Historically, the basin's socio-economic system has been based on livestock production which helped define its political and economic interests, and its main social and cultural characteristics. In the lower sub-basin, near Corumbá, the Urucum manganese reserves are estimated at 100 million tons. In the same location, iron ore reserves are estimated at 800 million tons. Elsewhere in the basin, known copper occurrences have good potential for exploitation, as do peat, lignite, gypsum, sapphires, amethysts, and topaz deposits. Minerals presently being exploited include gold, diamonds, limestone, marble, and clay.

Since the mid-1970s, the traditional balance between agriculture and mining has been upset due to the expansion of the agricultural frontier in the upper sub-basin. The principal factors causing environmental problems in this sub-basin are soil erosion, caused mainly by the large-scale, mechanized production of soybeans and rice, and water pollution caused by the intensive use of agrochemicals which are washed into the standing waters of the lower sub-basin. Soil erosion rates have been estimated at 300 tons/km^2/year in the upper sub-basin and at 40 tons/km^2/year in the lower basin. This problem is exacerbated by the compaction of the soil by heavy farm machinery, and the resulting lack of percolation of stormwater. The Pantanal is also being affected by land clearing in the areas of inflow and along river banks, by pollution from ever-growing agribusiness, and by uncontrolled urban and industrial discharges into the river system. Throughout the basin, fish are threatened by overfishing and, most recently, by the dumping of hazardous chemicals, especially large quantities of mercury used in gold mining.

Associated with these existing problems are the potential effects of specific development projects, including the highly polemical Paraguay-Paraná Waterway Project (Hidrovia Paraguay-Paraná, or simply Hidrovia), which seeks to facilitate the transport of iron and manganese ore from Urucum and Corumbá and the transport of agricultural products from Mato Grosso and Rondônia (located outside of the UPRB) through the Paraguay river system. In the latter case, some 5 million tons of grain are expected to be transported each year. It is necessary to quantify possible negative influences caused by the rectification and dredging of river stretches in specific areas, such as Posto de São Francisco and Fecho dos Morros. In these areas the river has a reservoir-like character which influences both the upstream hydrological behaviour of the UPRB and the downstream passage of flood peaks.

A study for the Integrated Development of the Upper Paraguay River Basin (EDIBAP) was conducted by the Government of Brazil, with the support of the Organization of American States (OAS) as executing agency for the United Nations Development Program (UNDP) between 1978 and 1981 (figure 4). It formulated a series of proposals with the objective of developing the Pantanal region based on a strategy founded on principles of environmental conservation, ecological balance, and rational land use.

Subsequently, in 1991, the Government of Brazil and the World Bank initiated the Pantanal Project, in collaboration with the Ministério do Meio Ambiente dos Recursos Hídricos e da Amazônia Legal (MMA), the Secretariat of Environment of the State of Mato Grosso, and the Secretariat of Environment and Sustainable Development of the State of Mato Grosso do Sul. The objective of the Pantanal Project was to establish measures to ensure the conservation of the Pantanal through the formulation and implementation of the Upper Paraguay River Basin Conservation Plan (PCBAP) and emergency actions for environmental protection. The PCBAP included the preparation of an environmental zoning study which contained general and specific guidelines for conservation, rehabilitation, and preservation of the natural resources; a geographical database for both states to make available physical, biological, social, legal, and economic information; and a real-time flood model designed to prevent negative impacts in urban and rural areas. Some emergency actions identified were the inspection and licensing of polluting activities, the inspection and control of exploitation of flora and fauna, monitoring of water quality in the basin, management and control of mining areas, rehabilitation of degraded areas, creation of a centre for the rehabilitation of wildlife, and promotion of informal environmental education activities.

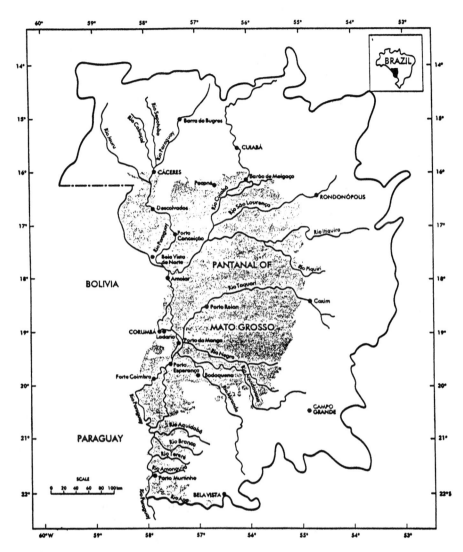

Fig. 4 **The Upper Paraguay river basin and Pantanal of Mato Grosso (Edibap, 1979; Project RADAMBRASIL, 1982a,b)**

The Pilcomayo river basin

The Pilcomayo river basin is one of the most complex sub-basins of the Plata system covering 272,000 km², or about 8.4 per cent of the Plata basin (figure 5). It is bound on the west by the Bolivian Andes, on the south by the Bermejo river basin, on the north by the Amazon

159

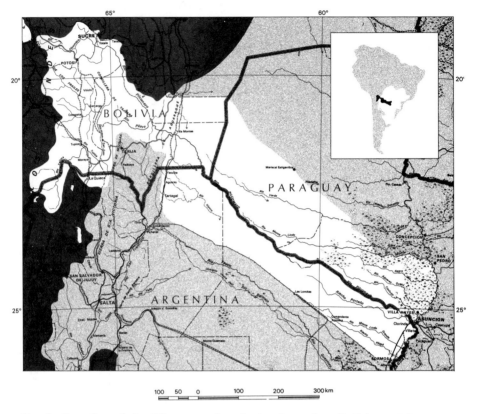

Fig. 5 **Location of the Pilcomayo river basin, Argentina, Bolivia, and Paraguay (OAS, 1984)**

river basin and portions of the Paraguayan Chaco, and on the East by the Paraguay river. The area is shared by Bolivia and Paraguay, approximately 35 per cent each, and by Argentina, with 30 per cent.

The two major divisions are the upper basin, which lies almost completely in Bolivian territory and ranges from 5,700 to 400 metres above sea level and, the Chaco region, an extensive plain that slopes smoothly to the east towards the Paraguay river. For the last 40 kilometres in the upper basin, the Pilcomayo river serves as the boundary between Argentina and Bolivia. Then, from Hito Esmeralda, where the boundaries of the three countries intersect, it forms the border between Argentina and Paraguay.

The Pilcomayo river discharge is very variable. During the rainy season, from December to April, it reaches up to 7,000 m^3/sec, while in the dry season, May to November, it stays below 30 m³/sec. The estimated 90 million tons of sediment that the Upper Pilcomayo annually deposits in the Chaco plain comes from geological and man-caused erosion in the upper basin, creating an extensive wetland (estero Patiño) some 280 km from Asunción. The estero Patiño wetland generates several small rivers and streams which cover a vast area, from the Verde river near Concepción (Paraguay) to the Porteño Creek near Formosa (Argentina), creating the interior delta of the Pilcomayo. In the 1940s, the Pilcomayo river began to sediment its own channel, a process up to now uninterrupted. Between 1968 to 1992 the point where the river deposited the sediment in the river channel receded more than 150 kilometres up river, at the rate of 7 to 10 km/year. By the end of 1996 only 60 km separated the point where the river still flows in its banks to Hito Esmeralda. Three hundred km downstream of the tripartite point the Pilcomayo river reappears. Fed by local rainfall and downstream groundwater discharge in a pattern unrelated to the surface hydrology of the Upper Pilcomayo river, the Lower Pilcomayo flows into the Paraguay river about 10 kilometres downstream of Asunción, Paraguay.

The basin's total population is 1.4 million, some 950,000 of whom are concentrated in Bolivia. The relative low average population density of 7.4 inhabitants per km² in this area is misleading, however, since nearly everyone lives in the steep and narrow agricultural valleys or in two important Bolivian cities – Sucre and Potosi – in the north eastern part of the basin. Outside these urban centres, vast areas are empty. In the lower basin, population density decreases to less than three inhabitants per km² in Argentina and to less than one in Paraguay.

Despite these caveats agriculture, cattle production, and population centres have considerable potential for expansion. All three countries have millions of hectares of agricultural and grazing land with high production potential, which could use regulated river water for irrigation and thus help invigorate the economies of the region and the three nations.

Studies conducted by the three countries in the mid-1970s and 1980s with support from the Organization of American States (OAS), indicated the need to identify and prioritize water projects in order to prevent the extensive upstream flooding by the Pilcomayo river. In addition to formulate short, medium, and long-term programmes and projects for the regulation of the river's flow so as to provide safe drinking water for the inhabitants of agricultural settlements building drainage, irrigation, and hydropower facilities. However, the studies recommended that precise topography fieldwork should be carried out and detailed river morphology investigations made before embarking on reservoir construction.

In order to deal with the problems of channel recession, some emergency measures were taken by Argentina and Paraguay in one of the critical points of the river, in an attempt to prevent the Pilcomayo river from entering Argentina. A joint request was presented by Argentina and Paraguay to the European Union (EU) to assist the governments to find a solution to the problem of retrogression and flooding in the Chaco. A mission of the EU visited the area in 1993 and endorsed the technical proposal prepared by both countries, even though the measures were not a permanent solution.

A definitive solution will only be possible with the construction of regulating waterworks in the Upper Pilcomayo in Bolivian territory where the necessary topographical and geological conditions exist to build reservoirs with adequate storage capacity so as to regulate river discharge and control the sediment load. The magnitude of the studies and the necessary investments make the participation of the three countries imperative, from the initiation of the studies through the stages of financing and construction of the reservoirs.

On 9 February 1995, the Governments of Argentina, Bolivia, and Paraguay, recognizing the necessity of establishing a permanent technical-juridical mechanism with which to manage the Pilcomayo river basin, created the Trinational Commission for the Development of the Pilcomayo River Basin. The main objectives of the Commission are to foster the sustainable development of the basin and its area of influence; to optimize the exploitation of its natural resources; to generate employment; to attract investments; and to allow the rational and equitable management of its water resources.

The Bermejo river basin

The Bermejo river basin is shared by Argentina and Bolivia. The impact of environmental problems that occur in the Bermejo basin is felt throughout the Plata system from the Andes to the coastal zone, and this influences many other developmental activities throughout the Mercosur countries, including impacts on the Hidrovia Paraguay-Paraná, the port of Buenos Aires, and the Plata estuary. Extending through the Tropic of Capricorn, the Bermejo river basin covers 190,000 km^2, 93.7 per cent in Argentina and 6.3 per cent in

Bolivia. It is approximately the size of the Rhine basin, and has a length of 1,200 km.

The Bermejo is unique in that it is the only river that crosses the huge area of the Chaco plains. Other major rivers in the region, such as the Timani, the Pilcomayo, and the Juramento, flow into the groundwater system of the plains and do not maintain their identity as surface water systems.

Being a continuous course of water, the Bermejo acts as a corridor, allowing the connection of biotic elements of both the Andean and Chaco plains ecosystems. Erosion and sedimentation are serious issues; it has been estimated recently that the Bermejo basin is the source of about 80 per cent of the suspended sediment present in the Plata river.

The present level of degradation of natural resources (both severe soil erosion and desertification) in the lower as well as in the upper basin, results in low levels of productivity of lands. Population in the basin is estimated at 1.2 million, a large part of it indigenous. Low levels of income force temporary migration of many local farmers to seek additional income, and this results in the general neglect of farmed land. Some survive through subsistence hunting and fishing, and others supplement their incomes selling regional handcrafts. Poverty and the low level of education are obstacles to any proposal for changes in the management of the basin. Under the present subsistence systems of production, simultaneous attention to economic profitability and environmental protection is difficult to achieve.

Although the area has been studied for many years (OAS Plata Basin Program, 1971–1973; Study of the Lower Bermejo, 1973–1975), it is only recently that action has been taken to implement major development projects in this basin. For example, in the upper basin, Argentina and Bolivia have agreed on the construction of a series of multipurpose water-resources development projects related to the general development of the region that could have potential impact on the downstream biomes. Programming the economic and social development of the region in a careful and orderly fashion leading to sustainable development is a challenge clearly recognized by both governments. For such purposes, a Binational Commission for the Development of the Upper Bermejo and Grande de Tarija River (a tributary of the Bermejo in Bolivia) Basins was created by the treaty of 9 June 1995. This Binational Commission has international legal status, full authority in technical, administrative, and financial matters, and the legal capacity to acquire rights and assume obligations. The Binational Commission is totally financed by, and acts on behalf of, the Governments of Argentina and Bolivia, and has been given authority by both governments to actively pursue all the actions required for the implementation of the present programme.

In October 1996, the Council of the Global Environment Facility (GEF) approved a request of the Governments of Argentina and Bolivia for assistance in the formulation of a Strategic Action Program for the Bermejo River Basin. The GEF-assisted project is being formulated at a time when the Binational Commission is considering the construction of several multipurpose dams along the binational border. These dams will change the present flow dynamics of the Bermejo river, creating opportunities for urban development and infrastructure, and agricultural development primarily downstream in Argentina. Anticipating and mitigating the impacts of these changes on the Bermejo River Binational Basin in a holistic manner, beyond the minimum requirements for environmental impact assessment, should be an integral part of the proposed SAP planning process.

The Gran Chaco

The Gran Chaco is a semi-arid ecosystem of 710,000 km^3 comprising parts of Argentina, Bolivia, Brazil, and Paraguay. Agricultural activities are expanding in the ecosystem, particularly in Paraguay and Bolivia. It contains areas degraded by land clearing, selective extraction of forest species, and erosion. There are areas of better soils and adequate water resource in which semi-intensive agricultural development is possible with the addition of improved pastures. Good examples are the Mennonite settlements in the centre of the Paraguayan Chaco (Pilcomayo river basin) and the area around Presidente Roque Sáenz Peña in the Argentinean Chaco (Bermejo river basin). Similar situations occur in the Argentinian and Bolivian piedmont. The two major rivers traversing the region – the Bermejo in its entirety and the Pilcomayo – have significant potential for supplying water for irrigation and the development of animal husbandry. In absorbing water and sediments, the Chaco system performs an invaluable flood and sediment-control function that could be matched only by the expenditure of many millions of dollars in infrastructure projects and dredging. How such systems are conserved could be critical to the development of the Gran Chaco region.

While of marginal quality, the forests of the Chaco contain species that are valued as timber (cedar and guayacan) and as sources of chemicals (quebracho and palo santo). Medicinal plants are often important sources of income, and the pharmaceutical industry is always interested in identifying useful species (about 250 species in Paraguay alone are in use). Some forms of terrestrial, aquatic, and recreational wildlife have potential for commercial development through the species' management and breeding (the cayman, river turtle, and capybara) for recreational hunting, and as sources of food for the local table. The germplasm is adapted to the temperature, drought, flood, and high salinity conditions of the Chaco soils, and could help improve crop production in other areas having similar conditions. There is great potential for research in biodiversity and its application. More than 320,000 hectares have been set aside in Argentina and Paraguay for five national parks and one research reserve.

The Laguna Mirim basin[1]

The Laguna Mirim is a binational water body between Brazil and Uruguay with a basin characterized by a dense network of rivers and small streams, as well as by low velocity flows among areas of high quality wetlands. The Uruguayan Eastern Wetlands, on the Atlantic coast, constitute the main ecosystem in the region for migratory birds. It stretches parallel to the coast as an extensive complex of wetlands, bogs, and lagoons draining into the Mirim lagoon. The Eastern Wetlands have been designated as a MAB Biosphere Reserve (1976) and a Ramsar site (1982).

The Laguna Mirim basin, covering an area of 62,250 km^2, of which 33,000 km^2 lies in Uruguay and 29,250 km^2 in Brazil, has undergone successive alterations. However, these have lacked a consistent focus and shown a strong pro-production bias in which important environmental factors have not been considered. Since March 1993, the Government of Uruguay has been implementing, with the support of the Global Environmental Facility (GEF), the "Program for Biodiversity Conservation and Sustainable Development in the Eastern Wetlands (PROBIDES)." Its main objectives are to achieve sustainable development in the region and launch a regional wetlands management system. The basin's main environmental problems are conflicts of use between the high productivity and biodiversity of its fertile wetlands, and the "reclamation" pressures from crop and livestock producers for construction of dams and drainage works. The development of rice-growing and the alternative uses of land for rice and cattle production have increased the pressure for drainage, which has received strong support from the private sector and the governments in the form of infrastructure projects. These conflicts have intensified with the use of the required fertilizers and pesticides for rice production. Moreover, the basin still remains at risk from thermal pollution from the Candiota thermoelectric power plant. This risk could be heightened by a proposed enlargement of that facility.

Conclusions

In view of the Plata's present and potential environmental problems, a strategy for the basin's sustainable development must include *a management plan for the environmental impacts resulting from development.* This will require the compilation, organization, and analysis of relevant data and information; rehabilitation and/or mitigation actions in specific areas or sub-basins; analysis of population dynamics, promotion of environmental education and public awareness; and fostering of public participation.

Planning and execution of development activities in specific areas, including the design and execution of specific investment projects, should be prioritized. These activities should seek to maximize the capacity for sustainable production and productive work; to improve the quality of life and standard of living of the inhabitants; to ensure conservation of the natural resources base, and to minimize conflicts over the use of natural resources.

Institutional strengthening and development and greater coordination of national organizations are urgently needed for these purposes. Strengthening and better organizing institutions in the basin, and providing them with adequate mandates, would foster coordination and the preparation of programmes and projects for the development and conservation of shared resources, the joint evaluation of environmental impacts of binational and multinational development projects, and the coordinated management of natural resources and shared ecosystems.

Transboundary water resources are playing an increasingly critical role in sustainable development as the economic and physical integration of the Plata countries proceeds. Plata basin countries should therefore try to identify and implement development projects which are strategically located and of general interest to the basin as a whole. *Plans are particularly needed for ecosystems shared by two or more countries* with a goal of furthering the integration of those countries. A sustainable development strategy for border areas should include (a) diagnostic studies of each area; (b) environmental zoning proposals that define areas suitable for sustainable production as well as areas for environmental protection; (c) integrated programmes that form part of an overall development strategy; and (d) national and binational investment projects formulated at the pre-feasibility and feasibility levels. These studies should also include the analysis and formulation of proposals of rational use of shared resources, facilitation of cross-border trade, and the possible integration and shared use of infrastructure and services in the areas of health, transportation, communications, and energy.

However, multinational development efforts have few chances of being successful without *a convergence of interests among the countries involved.* To this end, each country must clearly define its objectives at the beginning of the project negotiations and determine the advantages of the steps to be taken, including the social, financial, political, and cultural costs it is prepared to bear to implement the joint initiative.

Challenges and opportunities for regional cooperation

While all the countries of the Plata basin have recognized the need for, and expressed a commitment to cooperate in, the integrated development and management of the region, unilateral actions continue to prevail. An example of this is the preparation of national plans to evaluate the water and land potential for agricultural development within each country, that with exception of the Bermejo river basin, are being undertaken with minimum or no coordination or consultation between them. The Plata basin continues to have an enormous development potential, and conditions are better than ever for a cooperative and coordinated action to meet the needs and requirements of its population, while maintaining the required level of environmental stability and integrity of the basin.

The information gathered by the planning studies conducted in the Upper Paraguay, Pilcomayo and Bermejo river basins and the Laguna Mirim basin makes possible the implementation of developmental activity, particularly related to water-resources management programmes, soil erosion, and water pollution control, and the conservation and protection of aquatic ecosystems.

167

When planning for these areas countries should pay special attention in seeking to:

- *Promote and improve stakeholder participation* for water-resource management in the basin;
- *Strengthen basin institutions and build capacity* to enhance the ability of federal, state (province), and local agencies to manage the basin, including development of appropriate economic instruments to enhance the management of natural resources, optimization of the hydrological and sediment-monitoring networks, enhancement of the capacity to implement statutory permitting/ licensing requirements, the exchange and use of information and data between agencies and organizations, and the promotion of institutional networking.
- *Consider sub-regions as the basic units for planning,* especially at the level of ecosystems, and for regulation of natural resource use and, environmental conservation.
- *Link integrated water-resources management programmes* to social and economic development and address land and water uses and biodiversity conservation within the context of river basins and aquifers.
- *Take into consideration the aspirations and demands of local communities* as a fundamental motivating factor for setting priorities and social participation.

Actions or initiatives that could be undertaken by the Plata countries, include:

- *Facilitate access to existing information* on the severity and extent of the existing environmental problems in order to identify areas and sectors in which specific measures will be required.
- *Prepare an inventory of specific resources and ecosystems,* including the formulation of plans to manage in an integrated fashion these resources and ecosystems. The Plata basin contains regions and river basins which are a most appropriate geographic unit for planning and managing water rescues.
- *Identify, formulate, and implement transboundary water management projects.* In order to accomplish this there is a need for each riparian state to formulate water-resources strategies, prepare assessments, formulate action plans, establish priorities for biodiversity conservation and water-management activities, and harmonize those strategies and action programmes with the national policies and programmes of the other countries involved in order

to maximize the benefits of resource development. Cooperation between riparian states of the Plata basin should be implemented in conformity with existing agreements either for specific ecosystems or sub-basins.

At the sub-basin level, some specific actions and initiatives that could be undertaken:

The Upper Paraguay river basin
– *Clarify* the role of all existing stakeholders in the UPRB and, where appropriate, propose new role to ensure proper integrated water-resources management;
– *Define and evaluate* the nature of interactions between the UPRB, the Pantanal region and the Lower Paraguay river, under various development scenarios;
– *Identify and promote* the use of economic instruments to support water-resources management measures in the UPRB;
– *Rehabilitate* selected degraded areas as pilot demonstration projects to gain information for management purposes. These areas would be representative of the major classification of problems including mine waste mitigation, watershed erosion, and non-point pollution sources.

The Pilcomayo river basin
– *Thoroughly understand* the impact of the upper basin structural modifications on the floodplains, the eventual increased erosion of the river channel downstream from the proposed dams, the impact on aquatic and forest ecosystems and other environmental components and processes, before dam construction begins.

The Bermejo river basin
– *Formulate* a Binational Strategic Action Program (SAP) for the basin that addresses the resolution of priority transboundary environmental issues as part of the implementation of the water-resource and environmental agreement between the countries;
– *Incorporate* transboundary environmental concerns, including biodiversity and land degradation protection into the development policies, plans, and programmes for the basin in Argentina and Bolivia;
– *Establish a system of public consultation* on the implementation and development of sound projects for both countries.

The Laguna Mirim basin

– *Implement* the measures identified by the Joint Mirim Lake Commission (Comisão da Lagoa Miriam Brazil-Uruguay (CLM)) for developing agriculture in the higher areas of the basin that are best suited for crop production.
– *Consider*, in order to relieve the existing land pressures of rice growers and protect the wetlands other measures including damming the Jaguarão river to supply water for irrigated agriculture.
– *Increase* the monitoring of air, soil, and water quality in order to acquire objective data. Develop mechanisms to manage irrigation systems so as to allow the protection and conservation of specific areas and resources.

Notes

1. The Mirim Lake basin is not physically part of the Plata basin, but lies within its area of immediate influence, and should be regarded as such for purposes of integrated development or border-area integration.

References

Biswas, A.K., Jellali, M., Stout, G. 1993. *Water for Sustainable Development in the Twenty-first Century*, Water Resources Management Series 1. Delhi: Oxford University Press.

Biswas, A.K. 1996. "Water for the Developing World in the 21st Century: Issues and Implications." *Hydraulic Engineering in Mexico* (in Spanish). 11(3): 5–11.

Bozzano, B. and Weik J.H. 1992. *El Avance de la Deforestación y el Impacto Económico. Proyecto de Planificación del Manejo de Recursos Naturales.* Asunción: MAG/GP-GTZ.

Cordeiro, N.V. 1993. *Possíveis Projetos e Ações para o Desenvovimento Sustentável na Bacia do Prata.* I Simpósio Internacional sobre Aspectos Ambientais da Bacia do Prata. Brasil: Instituto Acqua, Foz do Iguaçu.

—— 1994. *Management of International River Basins.* II Congreso Latinoamericano de Manejo Integrado de Cuencas. Oficina Regional de FAO para America Latina y el Caribe. Merida, Venezuela.

—— 1996. *Integrated Management of Water Resources and River Basins.* Address to II Inter-American Dialogue on Water Resources Management. Buenos Aires, Argentina.

Dugan P.J. 1990. *Wetland Conservation: A Review of Current Issues and Required Actions.* Gland, Switzerland: IUCN.

Easter, K.W., Feder, G., Le Moigne, G., and Duda, A. 1993. *Water Resources Management*, World Bank Policy Paper. Washington, D.C.

Le Moigne, G., Barghouti, S., Feder, G., Garbus, and L. Mei Xie. 1992. *Country Experiences with Water Resources Management: Economics, Institutional, Techno-*

logical and Environmental Issues, World Bank Technical Paper No. 175. Washington, D.C.

OAS Plata Basin Program: Study of Water Resources in the Upper Bermejo River Basin (Argentina and Bolivia), 1971–1973; *Study of the Lower Bermejo River Basin* (Argentina), 1973–1975.

OAS 1984. *Integrated Regional Development Planning: Guidelines and Case Studies from OAS Experiences – The Pilcomayo River Basin Study*. Washington, D.C.

OEA 1973. *Desarrollo de los Recursos Hídricos de la Alta Cuenca del Rio Bermejo Argentina and Bolivia*. Washington, D.C.

—— 1977. *Uso Múltiple de la Cuenca del Rio Pilcomayo – Argentina, Bolivia and Paraguay*. Washington, D.C.

—— 1981. *Desenvolvimento Integrado da Bacia do Alto Paraguay (EDIBAP)*. Washington, D.C.

—— 1985. *Desarrollo Regional Integrado del Chaco Paraguayo*. Washington, D.C.

—— 1991. *Estudio Ambiental Nacional y Plan de Acción Ambiental*. Oficina de Planeamiento y Presupuesto de la República Oriental del Uruguay, Organización de los Estados Americanos y Banco Interamericano de Desarrollo. Washington, D.C.

Ponce, V.M. 1995. *Hydrologic and Environmental Impact of the Parana-Paraguay Waterway on the Pantanal of Mato Grosso, Brasil – A Reference Study*. San Diego State University.

Programa Nacional do Meio Ambiente/Jorge Adámoli 1995. *Diagnóstico do Pantanal*. Brasília: DF.

Programa Nacional do Meio Ambiente 1995. *Plano de Conservação da Bacia do Alto Paraguai*. Brasília: DF.

Rebouças, A.C. 1996. *Situação Atual das Pesquisas do Aquífero Gigante do MERCOSUL*. Seminário Internacional Aquíferos do MERCOSUL. Brasil: Curitiba.

The World Bank 1990. *Brazil: National Environment Project*. Report No. 8146-BR. Washington, D.C.

Annex

Activities carried out in the Plata basin with the support of the Unit of Sustainable Development and Environment General Secretariat (OAS)

Plata basin programme

– Inventory of Hydrologic and Climatologic Data. 1967–1969 (S,P)
– Inventory and Analysis of Basic Information on Natural Resources. Index maps. 1968–1971 (S,P)
– Energy Potential and Infrastructure (Update). 1981–1983.
– International Seminar on the Gran Chaco. 1983 (S)
– Transportation in the Plata Basin (Update). 1983–1984. (S)

Joint studies by two or more countries

– The Bermejo River Basin: I – Upper Basin: Argentina and Bolivia. 1971–1973.
– Multiple Use of the Pilcomayo River Basin: Argentina, Bolivia and Paraguay. 1974–1977.

171

National projects and studies

Argentina

- The Bermejo River Basin: II – Lower Basin. 1973–1976.
- Environmental Quality and River Basin Development: A Model for Integrated Analysis and Planning. 1975–1978. (E,S)
- Integrated Development of the Province of Misiones. 1978–1979.
- Development of the Province of Formosa. 1979.

Bolivia

- The Bermejo River Basin: III – The Bolivian Region. 1975–1976.
- Investment Plan for Agricultural Development with Irrigation in the Upper Pilcomayo Basin. 1980–1982.
- Agro-industrial Projects in the Department of Chuquisaca. 1985–1987.
- Development of the Bolivian Chaco. 1987–1989; 1991–1994.

Brazil

- Northwestern Paraná State: I – Erosion Control. 1970–1972. (P)
- Northwestern Paraná State: II – Study for Regional Development. 1973–1974 (P)
- Integrated Development of the Upper Paraguay Basin (EDIBAP). 1977–1981. (P)

Paraguay

- Development of the Northeastern Region of Paraguay (Aquidabán Project). 1972–1974.
- Development of the Paraguayan Region of the Pilcomayo River Basin. 1980–1982.
- Integrated Regional Development of the Chaco. 1982–1985.
- Consolidation of the Rural Settlements of Coronel Oviedo-Mbuty and of the Concepción-Pedro Juan Caballero Axis: Feasibility Studies. 1982–1984.
- Project for the Reconstruction and Integrated Development of Metropolitan Asunción and of the Flood-Stricken Cities of Concepción, Alberdi and Pilar. Report of the Preliminary Mission. Washington, 1992.

Uruguay

- The Santa Lucía River Basin: Water Resources Development. 1969–1970.
- Irrigation in Canelón Grande and Aguas Blancas: Prefeasibility Study. 1876–1978.
- Support to the Department of Regional Planning of the Secretariat for Planning, Coordination and Public Information (SEPLACODI). 1979–1981.
- Traffic and Transportation in the City of Montevideo. 1991–1992.
- National Environmental Study: Environmental Action Plan. 1990–1991.
- Regional Project of Alternatives for Forest Investment. 1992–1993.
- Transportation Alternatives for Uruguay in the Context of MERCOSUR. 1992–1996.

– Study of the Regional Integration of Freight Transport. 1994–1996.
– Regional Project on Alternatives for Forest Investment. Phase II. 1994–1996.

(S) Spanish
(E) English
(P) Portuguese
In the absence of any indication, the reports are only available in Spanish.

7

The Plata basin institutional framework

Lilian del Castillo Laborde

The first steps forward

The Plata basin as an organic internationalized structure is the outcome of several and confluent political conceptions which arose in
the second half of this century and took over international bodies
and governments alike. These ideas began to crystallize in the decade
of the 1950s, as technical and assistance organs within the United
Nations Organization and in the embryonic organization of western
Europe.

During the 1960s the concept of development was paramount.
Countries were classified as *developed* or *underdeveloped* according
to adopted standards of *per capita* gross domestic product, a measure
of the strength of an economy.

It is a useless and disappointing exercise at this time to examine the
adequacy of such a concept of development and the disregard of local
approaches that it involves. However, since then, there has been a
political perspective within international organizations and influential
governments considering *well-being* a pre-condition for peaceful
relations among nations.[1]

Going back to its origins, in November 1954, the IV Special Meeting of the Interamerican Economic and Social Council[2] analysed a

document prepared by the Economic Commission for Latin America and the Caribbean (ECLAC) dealing with the willingness to have a joint development programme for the Americas, the urgent need to plan national policies, to establish priorities in government expenses, to encourage a tax and agrarian reform, to promote international commerce and to enhance international assistance and cooperation. These were the first discussions that paved the way to constituting future regional institutions.

In September 1958 the Foreign Ministers of the 21 member countries of the Organization of American States (OAS) held an informal meeting in Washington, D.C., to consider the Panamerican Operation, a Brazilian-based proposal which strengthened improvements for Latin American states in economic integration, transport and electric power growth, industrial diversification, commercial expansion, and a massive educational push-forward. To follow up these goals, a 21-member Special Committee was appointed. On its first meeting, in November and December of that year, the commitment of the US Government to assist the programme of planned development financially stimulated the creation of the Interamerican Development Bank (IDB).

In the course of the next year, 1959, all basic agreements for the constitution of the Bank were concluded. In January 1960, the Governors of the IDB met for the first time, with a promised grant of $5,000m approved by the US Congress.[3] On 11 September 1960, the Special Committee enacted the Bogotá Act which stated the commitment of member states to reinforce social reforms in their countries.

To coordinate the planning of future tasks, a three-member Committee was also set up in 1960 with representatives of OAS, ECLAC, and the newly created IDB.

In August 1961 in Punta del Este, Uruguay, the Alliance for Progress was born to deepen economic and social development in Latin American states. It also aimed to support training and to finance studies, projects, and public works with a previous priority selection. To fulfil these objectives the Inter American Committee (CIAP) was organized. However, during the next decade and with uneven results, a not-so-substantial investment was channelled through to the Alliance members' beneficiaries.[4]

In 1965 in an effort to finance the economic growth of disadvantaged states and accelerate the flow of capital to developing countries, the United Nations decided to set up a new fund, the

United Nations Development Programme.[5] The new institution had in some respects concurrent functions with the UN Conference on Trade and Development (UNCTAD) established in 1964[6] with overlapping targets but different implementation capacities.

The institutions for supporting the Latin American economic transformation were founded. The scenery was set for master plans, feasibility studies, terms of reference, consultants, each of them dependent upon the technical and financial assistance derived from international bodies.

There was another concept suitable for development in that feverish decade of the 1960s, and that was the movement towards integration. Development would be better promoted jointly, not only to speed up the benefits for the region smoothly, but as an integral part of the process.

The political conditions of South America were nevertheless not encouraging for an integrative development. Perhaps some improvements were made in regional trade, as was the case with the Latin American Free Trade Association (ALALC) with some categories of product included and long lists of products excluded from beneficial treatment. However, there was no actual commitment to intra-regional relationships.

As development and integration conceptually melted into each other there was another requisite to make this model applicable, the geographical or cultural aspect of a particular region. For this function, catchment areas adapted perfectly as most populated and industrialized sectors are situated on the banks of main rivers. It was then considered especially convenient to design the support of economic growth on the given surface of a basin system, used in that context as a welding element of independent factors. The election did not change the fact that riparian states did not necessarily share, and often do not, any common goals. As benefits and penalties are both outcomes of the integration and development process, implicit in transformation, divergence is not an obstacle for agreement provided that common targets are intended and achieved.

From this standpoint it is useful to bear in mind that riparian states are not free partners but compulsory joint owners. The basin area has to become an integrated geographical unity, suited to carry out a joint development programme. Out of this conception, development, integration, international bodies, technical and financial assistance, and the Plata basin became a common draft.

Aiming towards a legal water regime

On 27 March 1963, the Brazilian representative before the OAS addressed a letter to the Secretary General suggesting a Panamerican conference on the exploitation of waters of international rivers with the purpose of stating the rights and duties of riparian states, to clarify their authority and capacity and to establish the basis of a convention which thoroughly regulated these topics.[7]

On 30 August 1963, the Interamerican Juridical Committee approved an outline called *The industrial and agricultural utilization of international rivers and lakes.* Pursuing the government's observations and a broad discussion of the text, the final draft convention was approved by the Committee on 1 September 1965.[8]

With a document ready for consideration, the Conference was convened by the II Extraordinary Panamerican Conference on 30 November 1965,[9] but it never took place.

Among other reasons, the Brazilian attitude towards the adoption of a convention on the industrial utilization of international waters experienced a drastic change. As stated in a Brazilian note to the Argentine Government on 29 January 1969, there could only be serious and deep disadvantages in trying to institutionalize what needs to be a natural and spontaneous cooperation, since every time a country chooses to reply to suggestions there has to be an exercise of full sovereignty.[10]

The consideration of an agreement on international rivers was deleted from the agenda of the Panamerican Operation. Since the possibility of approving general rules was out of the question, there remained the opportunity for neighbouring basin states to adopt special understandings and agreements.

In 1966, as the five countries of the Plata basin began to look on themselves as riparian states, there was a first approach to negotiating certain instruments, pointing out their common interest in the river system.

At the standpoint of the Plata basin structure, 30 years ago, each of the five countries of the region, Argentina, Bolivia, Brazil, Paraguay, and Uruguay had, and quite certainly they have today, contrasting approaches to regional development, emerging from each country's different historical, geographical, social, and political background. Nonetheless, this does not exclude the existence of common goals for member states. We can accept common goals which start from different, even opposite, perspectives. They are no more than dissimilar ways to reach the same objective. This means, in other words, a regular situation of confronting interests, a constant in international politics. Furthermore, and this was a positive point, there are no boundary issues in the Plata region, or at least no significant ones.

With the priorities of social and economic demands directing the urgent use of water resources towards the production of power supply but without an adequate water policy, either national or regional, there was little room for water planning objectives. At that time the main issues were the utilization of water slopes for hydro-electric power generation, with the consequence of the dams' impact on the river system, subsidiary attention to navigation, and little concern for water quality and other topics.

Previous to convening the Plata Basin Ministers of Foreign Affairs first meeting, the Government of Argentina informed the Interamerican Development Bank (IDB), in early 1966, of its aim to pursue an integral survey for regional benefit in the Plata basin system. On 6 June 1966, Argentine Foreign Minister Zavala Ortiz took the initiative for a conference to be held in Buenos Aires to reach an integral and coordinated plan for the utilization of the Plata basin.

The riparian states agreed at a subsequent conference, on 27 February 1967, to start a joint and comprehensive survey of the Plata basin with a view to carrying out a programme of works, either national, binational, or multinational, adequate for regional development. A basic international organization was established to undertake this mission, the Coordinating Intergovernmental Committee (CIC), to be located in Buenos Aires, Argentina.

From the beginning we were aware of two different conceptions, first the aim of Argentina and Uruguay towards a joint and integrated management of the basin and second, the list of works included in the Final Act of the meeting by Brazil, Bolivia, and Paraguay, either connected or not connected to the basin system. This first approach made it clear that for Argentina, and probably Uruguay, the Plata basin was mainly, though not exclusively, water management, while for the other riparians, it was mainly the development of the area bounded by the basin.

The outline of the Plata Basin Treaty

To follow up the decisions of the Buenos Aires Conference, the IDB convened in Washington, D.C. on 10 November 1967, a meeting of international organizations connected to the region in order to assist

179

the Plata basin states in their incipient coordinating efforts. It was agreed to constitute the Consultative and Coordinating Board for the Plata Basin Development Programme with the following members: (a) Institute for Latin American Integration (INTAL), created by the Executive Board of the IDB in 1964 and organized in 1965 as a special programme branch to study the process of integration in Latin America; (b) Interamerican Development Bank (IDB); (c) Organization of American States (OAS); (d) Interamerican Committee for the Alliance for Progress (CIAP); (e) United Nations Development Programme (UNDP); and (f) Economic Commission for Latin American and the Caribbean (ECLAC). The objective of the Consultative Board was to provide for the planning and financial aid necessary to accomplish the studies that would allow for future programmes and projects. INTAL was appointed as the Secretary of the Consultative Board to coordinate and manage the technical assistance supported by its members.[11]

The Consultative Board held meetings with experts of the five basin countries in November 1967 and in April 1968. At the same time the Plata Basin Committee, CIC, also held meetings in Buenos Aires to analyse the projects submitted by Bolivia, Brazil, and Paraguay.[12]

The financial assistance authorized by the IDB amounted to US$700,000 in June 1968 for promotion and coordination of tasks and basic studies to assist governments in their future projects. The terms of reference for these studies were approved. There was also the proposal to carry out technical studies for preferential areas and the offer to encourage governments in a long-term programme for the basin development.

As we have mentioned, the concept adopted by the Consultative Board did not fit in with the objective of water management. To repeat the OAS words, what was prepared was a joint planning action for the development of the dredge-bounded zone considered a geo-economic unity and embracing the territories of various countries.[13] The original proposal meant concrete goals related to the use of natural resources, navigation, and flood control,[14] thus becoming shorter in terms and larger in ambition.

On 18–20 May 1968, the Second Conference of Plata Basin Foreign Ministers took place in Santa Cruz de la Sierra, Bolivia, and a fundamental change occurred. On that occasion the earlier proposal of cooperative and coordinated water planning took a back seat. Instead, infrastructure priority projects of great importance came up, as listed in the Santa Cruz de la Sierra Act. The decision on water policy could be summed up as the maximum utilization of basin waters, picking up projects of maximum utility (as in *Considerations*, para. 3, and *Resolutions* C1. and 4 of the Santa Cruz de la Sierra Act). This and other criteria stated in the Act could be overlooked by governments' political reasons to impel a certain project (as stated in *Resolutions* C, final paragraph of the Act). The basic principles included in this Act would be the fundamental guidelines for the future Plata Basin Treaty, to be devised by CIC and approved by the Plata Basin Foreign Affairs Ministers. Nonetheless, the drawing up of a water management statute for the Plata Basin hydrological resources that was recommended in the same Act (see *Recommendations*, para. 1) is still pending.

This radical change in the original model and the absence of political coincidence by riparian states began the operational misunderstandings between intervening international organizations and provoked the collapse of the Consultative Board, which ended with the resignation of its Secretary in 1969.

On 22–23 April 1969, the First Extraordinary Conference, held simultaneously with the Third Conference of Plata Basin Foreign Affairs Ministers, convened in Brasília for the purpose of adopting the Treaty of the River Plate Basin,[15] was intended to "remain in force for an indefinite period" and was named that way after Article VII of the treaty.

The different aims of *development* (Article I, only paragraph, item (e), Article II, first part; Article III, first part), *integration* (Article I, only paragraph, items (a), (d), (f), and (g)), *water management* (Article I, only paragraph, items (b), (h), and (i)) and *environmental concern* (Article I, only paragraph, item (c)), seemingly compatible, proved not to be so when the system was put to the test.

The treaty, approved by the five member states and in force at present, incorporated as organs of the Plata Basin structure the Conference of Foreign Affairs Ministers and the existing Committee, CIC. The Statute of CIC, already adopted at the Santa Cruz de la Sierra II Conference, added a Secretariat to the Committee, to hold office in the Buenos Aires seat. The treaty institutionalized the precedent disconnected stages into an organic-shaped system.

Both instances, the Committee and the Foreign Affairs Ministers Conference, had a political structure. As no technical support was included, most certainly because the technical assistance of international bodies was anticipated,[16] technical groups of an advisory nature were incorporated.[17]

The sporadic work of those technical groups was later coordinated[18] and they became another step in the treaty's institutional system.

Complicated beginning

The confronting views stated in the Santa Cruz de la Sierra Act which lay in the Plata basin institutional origins have never been solved and, in the records of those initial meetings and its attendant documents, it is possible to trace the opposite views and interests of member states.

Big dams, large bridges, and navigational facilities, among other things, have been constructed and the preference for economic development has paved the way for specific institutions to share stretches and undertakings. But water management is still pending.

During this period of 30 years, as has been explained above, the population grew, the energy demand expanded, the expansion of the agricultural frontier leading to soil erosion and sedimentation remains as serious as ever. But water management is still pending.

At present, there is growing awareness about the increasing pollution of the catchment area, the persistence of flood threat, the increasing deforestation and the degradation of fertile land, and recent agreements have expressed the aftermath of environmental concern. But water management is still pending.

The Plata Basin Treaty (Article I, only paragraph, items (b), (h) and (i)) states a comprehensive regime to be one of its objectives. With international support organized and the Plata basin institutionalized riparian states had attained the most favourable structural conditions for making progress. The different conceptions of water administration and basic infrastructure works were implemented in different ways, though connected and in some circumstances cooperative.

The study of the basin itself was mainly carried out by OAS, in a formidable task based on data collection and field operations that has not been surpassed,[19] called *Inventory and Analysis about Basic Information on Río de la Plata Basin Natural Resources*, with the financial support of OAS, Development Assistance Special Fund. This survey included the hydrological and climatologic inventory, with sections on aerial photography, topographic and planimetry maps, geological maps, soil and land productivity maps, vegetation, ecology, land use, and forestal inventories maps.[20]

Many other important surveys were carried out by OAS, Natural

Resources Unity. We can mention, namely, the Study for the Control of Erosion in the Northwest Paraná State, Brazil, executed in the period 1972–1974,[21] the Aquidaban Project for the Paraguay North East Region, between 1972 and 1974,[22] the Study of the Hydrological Resources and Development Programming on the Upper Basin of the Bermejo River, Argentina and Bolivia, in the period 1970–1973,[23] the Development Programming for the Lower Basin of the Bermejo River, Argentina, in the period 1973–1976.[24] To reach a consensus on the standards to be applicable to water management was at the core of the Plata basin policy of both Argentina and Brazil.

In the meetings of CIC of 1970, Argentina and Brazil expressed their different interpretation of the Plata River Treaty, reflected in the mandate for the group of experts on water resources.[25] Argentina wished to draw a set of general rules applicable to the basin water resources. Brazil requested the acceptance of its own technical judgement as enough guarantee for other riparians in relation with existing and planned hydro-electric power plants. As stated by its representatives, it maintained that a country possessing the sources of a drainage basin could not willingly limit itself on the uses of the waters and the only acceptable restraints could be those arising from technical reasons and its principles of legal responsibilities.[26] It needs to be noticed that this position is incorporated in Article V of the Plata Basin Treaty, which states that "Any joint activities undertaken by the Contracting Parties shall be carried out without prejudice to such projects and undertakings as they may decide to execute within their respective territories, in accordance with respect for international law and fair practice among neighbouring friendly nations."[27]

There were understandable arguments for these claims. A large stretch of the Paraná river sub-basin, situated in Brazilian, Argentine and Paraguayan territories, has relevant conditions for hydroelectricity, such as appropriate slope, important flow, basaltic structure, and embanked stretches. The successive stretches of the Paraná river have different status, as it spills over Brazilian territory upper stream, then over the Paraguayan-Brazilian border, over the Paraguayan-Argentine boundary and finally, in its extensive lower stream, over Argentine land. Thus, the internationalized nature and valuable qualities of the river were able not only to generate hydroelectric power but also to raise bitter confrontation between riparian states. Since 1960, Brazil has launched the construction of numerous barrages in the basin, in a restless building effort which is still current and will extend into the future.[28] Paraguay and Argentina, lower riparians, were permitted to construct two important dams in their shared stretch at the same period in the 1970s.[29]

Uruguay is not a riparian state of the Paraná river and Paraguay adopted a waiting role. For the same reason, Argentina held its isolated position versus Brazil as regards the Paraná river and pressed the incorporation of general international law rules applicable to the uses of international water resources as suitable rules for Plata basin states.

In a conciliatory success the Plata Basin States Foreign Ministers established the basic principles for water management applicable to Plata basin riparian states in the *Asunción Declaration on the Uses of International Rivers*,[30] approved in June 1971. Those principles, general and limited, could be rephrased as follows:

(a) In contiguous rivers, as riparians share their sovereignty, every use of the watercourse should be preceded by bilateral agreement of riparian states.

(b) In successive international rivers, where riparians do not share their sovereignty, each state is able to use the watercourse according to its needs provided the uses thereof do not cause appreciable harm to another basin state.

(c) Riparian states agree to exchange hydrological and meteorological data and cartographic results from field measurements.

(d) There is an emphasis on the improvement of river navagability and a warning that future works should not hamper navigation.

(e) States are required to take into consideration the living resources of basin waters in works planning.

The distinction between contiguous and successive international rivers is restricted to the concept of rivers as parts of the basin, and it excludes tributaries and underground waters. It considers rivers in their particular stretches and disregards water resources and basins as unity entities. On the contrary, it stresses the discretionary possibilities of riparian states as regards the utilization of water resources in their own territories, though restricted to the obligation not to cause appreciable harm to other basin states. It was also clear that the statement incorporated as the rule applicable to successive rivers according to the Asunción Declaration is the rephrasal of Article V of the Plata Basin Treaty.

In the Asunción Declaration, then, there are two different rules incorporated as regards the contiguous and successive international watercourses. The norm of prior consent is applicable to the former, and the norm of previous consultation is applicable to the latter. Whereas the Asunción principles are constrained they were enough to bring objective standards into being and from that one circumstance many other agreements became possible for water undertakings in the basin.

Though not a treaty, the Asunción Declaration, as we have seen, expressly ruled that on behalf of riparian states the rule should not cause "considerable harm" in the utilization of international water resources. This, however, was not the only rule that lower riparians, mainly Argentina, wished to incorporate as mandatory principles for basin undertakings. It maintained that the principle of equitable and reasonable use of fresh water resources and the rule of previous consultation were also applicable as general international law rules regulating the use of international basins. The other riparians, especially Brazil, were prepared to allow only those restrictions incorporated by treaty for each particular use.

These different positions became an issue at the Stockholm Conference of the United Nations on the Environment, 1972, and the general principles on international shared resources were incorporated as General Assembly Resolutions 2995 (XXVII) and 2996 (XXVII) of that year, and were followed up by Resolutions 3129 (XXVIII) and 3281 (XXIX), whose Article 3 recognized the rule of previous consultation.

We can point out that the previous consultation principle is more a procedural than a substantive rule, and is not intended to invest other riparians with a veto capacity but to allow them a transactional opportunity. It regularly allows riparian states potentially affected by future works to enter into negotiations to make their requirements compatible. This does not mean, moreover, that the acceptance of a procedure of previous consultation would imply an unreasonable delay for the other riparian or even the obligation to reach an agreement. The scope of previous consultation practice was carefully considered by the award at arbitration in the Lake Lanoux controversy between France and Spain,[31] which recognized the features of this procedure. The procedure usually needs to be established by agreement of the parties and to be linked to a settlement of disputes mechanism to ease the process for the use of international waters and international resources.[32]

In order to make a qualitative improvement both on the legal nature and on the formulation of the Asunción Declaration, the rules about the scope and content of activities implied in the principle of "considerable harm" would be drawn up by the Plata basin riparian states. This development would in the end mean wading into the water-management field and enriching the elementary statements of this document.

After the adoption of the Declaration other treaties were made by basin states for subsequent water works in tributary basin rivers and in each case new commissions were established, but the inclusion of substantive rules on water management was in every case avoided. These rules would have incorporated: (a) the commitment of member states to monitor the quality of the water body, which consequently would imply accepting pollution as one of the causes of "harm," (b) monitoring the quality as well as the quantity of water entering the territory of lower riparians; (c) the attitude towards special situations such as river diversion, and (d) the many other cases to be considered, from the preservation of living resources to irrigation or transportation. These primary rules should be crowned with procedural rules on dispute settlement, thus incorporating a complete legal instrument into the basin system.

Water management considers not only the rights of riparian states of international catchment areas. It also clarifies the duties that states are reluctant to subscribe to. Nevertheless, it has to be admitted that law, whether national or international, is an indissoluble amalgamation of rights and duties, where the disregard of one's duties is also the disregard of somebody else's rights.

Uses of the watercourse

It is important to remember that the main uses of the Plata basin rivers are navigation and the generation of hydroelectric power. Freedom of navigation on the Paraná, Paraguay, and Uruguay rivers was granted to riparian and third states in the 1850s, both in treaties and national statutes.[33] The legal system established in the last century remains almost unchanged as the legal background of the Plata basin navigational law, with the addition of new regimes. The most significant at the international stage is the multinational Paraguay-Paraná waterway, from the town of Cáceres in Brazil to the New Palmira harbour in Uruguay.

As happens in other international river basins in South America, there is no general customary rule to establish freedom of navigation in the Plata basin. This right is conventionally granted to riparian states by means of treaties or enacted by national legislation.[34]

The navigable network of the basin extends for some 8,600 kilometres. As some of the tributaries of the basin carry large volumes of silt caused by soil erosion, there is the expensive effort of dredging for riparian states to maintain navigation channels in proper conditions. The Río de la Plata navigable channels, the access to the Plata basin fluvial system, have been built and preserved by Argentina, which also covers buoying in the river.

Works for hydroelectric power generation are the main undertakings in the Plata basin. Besides the important amount of national work carried out by Brazil, already quoted, the favourable outcome of the Asunción Declaration and the possibilities of financial assistance by international bodies which became generally available in the seventies encouraged riparian states to enter into bilateral agreements for construction of the most important projects.

On 16 June 1971 Argentina and Paraguay entered into an agreement to research the utilization of the Paraná river in their common stretch between the Iguazú and the Paraguay rivers, and to that end the Argentine-Paraguayan Mixed Commission of the Paraná River was founded. The commission came into force on 15 March 1972. Both countries identified the Corpus Christi location, in the stretch where the rivers run as a canyon, as the site of a binational enterprise, the Corpus barrage.

Almost coincidently, on 26 April 1973, Brazil and Paraguay signed the treaty to develop Itaipú, upstream to Corpus and in the same stretch of the Paraná river canyon. On 3 December 1973, Argentina and Paraguay subscribed to a similar treaty to develop Yacyretá, quite close and downstream as regards Itaipú and the Corpus works, on the Paraná canyon. In order to draft, built and later operate these enterprises binational organizations were established by the same treaties.

Meanwhile on the Uruguay river the Salto Grande project, which was agreed upon between Argentina and Uruguay in 1946 but had never developed further, became a fact with the signature of additional documents on 20 October 1972 and 12 February 1974.[35] The Salto Grande Mixed Technical Commission took care of the construction of this most efficient waterwork and is at present in charge of its operation. The Salto Grande dam was the first binational enterprise of the Plata basin riparian states. The first turbine entered into operation in 1979 and the last one was launched in December 1982, together with the international highway and railroad along the crest of the barrage. It is both an economically profitable and environmentally sound undertaking.

As has already been already mentioned, the Paraná river runs in one of its stretches in a basaltic gorge which begins at an altitude of 222.25 metres in Brazilian territory and ends in one of 73.54 metres, sea level, in the Argentine city of Posadas. At this point the river has a medium annual flow of 11.800 m³/sec., ranging from a maximum flow of 52.600 m³/sec. experienced in 1905 to a minimum flow of 2.900 m³/sec. measured in 1944. Which of the water fall or flow is the most important element for power generation capacity, determines the electricity potential of the watercourse. Thus, the geographically exceptional conditions of the Paraná river favoured the construction of a great number of dams in Brazilian territory. Additionally, three main binational hydroelectric works, the Itaipú, Corpus, and Yacyretá dams, though designed many years ago, were drafted in the stretch where the Plata basin started.

On 23 September 1960, Argentina, Paraguay, and Brazil signed a Joint Declaration about the possible consequences of the construction of the Yacyretá barrage on Brazilian territory.[36] As regards the Corpus and Itaipú projects, it was necessary to be aware of the characteristics of the Itaipú dam, namely the Itaipú operational regime, the restitution level of the Itaipú turbinated waters and the Corpus reservoir water-mark, when designing the Corpus to work at its utmost capacity. The unhelpful attitude shown by Brazil and Paraguay on forwarding the outlines of the Itaipú project to Argentina constituted a long-standing dispute of 12 years between these countries about the uses of the water body of the Paraná river.

The legal issues involved concerned the existence of customary international law rules mandatory between riparian states on international basins, as to the water-mark a state is entitled to use. The lower the restitution level of the upper stream dam, the lower the permitted water mark of the downstream reservoir, which means less profitable energy generation capacity. For the Corpus project in particular, each metre of difference in the reservoir water mark would represent US$32m a year, calculated at then current rates.

During the period 1977–1979 negotiations took place between representatives and experts from Argentina, Brazil, and Paraguay purporting to find out the compatibility and acceptability of an equitable solution. The outcome of the discussions was the Tripartite Agreement of 19 October 1979.[37] On its terms, the water-mark of the Corpus reservoir should not exceed the 105 m sea level and the Itaipú dam would operate with 18 turbines for a maximum water flow of

12,600 m³/sec. According to the treaty, the water-mark of the Paraná river must not fluctuate more than 0.50 m by the hour or more than 2 m by the day.

The agreement was the starting point of a cooperative policy as to the Paraná river power works by the three riparian states. The Itaipú dam is now [at the end of 1996] in full operation, the one of Yacyretá has reached half its capacity and the Corpus project, with an updated reformulation that diminishes its power generation capacity yet simultaneously reduces the reservoir surface and consequently the land flooding, will be ready for international bidding in 1997.

Once the compatible approach was achieved, a new flow, that of regional agreements, began. Receiving inspiration from the Tripartite Agreement, a treaty for the Uruguay river was agreed by Argentina and Brazil for the boundary stretch between them on 15 May 1980. From its clauses another binational undertaking, the Garabí barrage, has since been drafted and it is at present at the stage of executive project. As in the case of Corpus, the work is being reshaped for accurate cost-effective results.

In the nineties new and already existing developments are shifting from a state-owned entitiy perspective into a competitive private investment perspective. From this standpoint, less profitable though sometimes politically attractive projects demand to be rethought and redrafted. This is in response to three main factors. One of them is the transformation of regional state-controlled economies into an increasingly market-regulated frame, with the consequent demands on private investment, risk capital, and profitable return. Another is political confidence in the convenience of intraregional economic integration as a most welcome tool for overcoming both the traditional extraregional dependent trade and helping adjacent countries to attain more focused relationships. The other factor is the incorporation of environmental patterns, emerging in the international field from Stockholm 1972 to Río 1992, and gaining enforcement afterwards mainly by the constraints of international financial bodies, as is the case with the World Bank.

These coincident structural and attitudinal changes have brought about a redefinition of many infrastructure projects and a redesigning of those already drafted, in order to incorporate the new trends. As an example, the new site for the Corpus work will be Itacurubí, in the Argentine-Paraguayan stretch of the Paraná canyon, 44 km upstream in the Paraná river from its former site in Itacuá. In this location the environmental impact will be minimized, diminishing the flood land surface by 50 per cent from 20,000 to 10,000 hectares[38] with a similar generation of 20,000 Gw/h yearly.

The satellite organization

The Plata Basin Treaty is broadly comprehensive as regards its competence on plans, projects, works, and programmes in the catchment area. Nevertheless, it is not proposed as an exclusive option for riparian states, but as a framework agreement that could add special benefits to its global scheme.

Accordingly, Article VI, which states that "The provisions of this Treaty shall not prevent the Contracting Parties from concluding specific or partial bilateral or multilateral agreements designed to achieve the general objectives of the development of the Basin," could be admittedly considered as a reservation as regards the scope of the treaty itself and as interpretation of the wording of Article I, only paragraph, when it refers to "operating arrangements and legal instruments." The instruments to be achieved by member states then, could also include those agreements envisaged by Article VI. Soon, after the basic agreement on the rules applicable to water management appeared in the Asunción Declaration of 1971, riparian states started to engage in specific and mainly bilateral agreements and projects.

On 19 November 1973 in Montevideo Argentina and Uruguay signed the Treaty on the Río de la Plata and the corresponding Maritime Boundary.[39] This treaty settled the controversial situation about the exercise of jurisdiction over the vast river's waters. Apart from jurisdictional matters, the treaty deals with navigation, fishing, bed and subsoil, pollution prevention, pilotage, works, scientific research, and rescue operations, among other aspects of the river system.[40] It also set up two permanent commissions, the Administrative Commission of the Río de la Plata and the Joint Technical Commission, for the adjacent maritime zone and the overlapping common fishing zone.

The Rio de la Plata Administrative Commission, established by the treaty and settled by complementary agreement of 15 July 1974, has its official location in the Argentine Martín García island, a strategic site in the Plata river which could control the access to the Paraná and Uruguay rivers and the object of a thorny dispute as regards

Argentina and Uruguay, both claiming historic and legal titles for its possession.[41]

The commission has management and administrative functions. In addition to its research programmes, mainly on silt, fisheries, fluvial tides, and water quality, it has been appointed by member states to design, bid, and adjudicate upon the dredging and buoying of the Martín García navigation channel.[42] This work, to be carried out in 1997, will be the first navigation channel of the Río de la Plata to be built and administered by both riparian countries. The existing channels were made, maintained, and managed by Argentina, as the riparian state of the Plata and Paraná river harbours for navigation channels.

The Joint Technical Commission for the maritime zone adjacent to the Río de la Plata is located in Montevideo, Uruguay, and has been working since January 1976. It has comprehensive capacity as regards living resources and the marine environment of the common fishing zone, namely, to establish the catch volumes per species, to promote joint scientific studies, make recommendations for ensuring the value and balance of bio-ecological systems, draw up plans for the preservation and development of living resources, establish standards for the reasonable exploitation of species in the common zone and the elimination of pollution.[43] The most economically important fish stock in the common zone is hake (*merluccius merluccius hubbsi*), not limited to this zone but sprawling over the larger part of the Argentine exclusive economic zone.[44]

On 26 February 1975, Argentina and Uruguay agreed on the establishment of a special body for their shared stretch of the Uruguay river, between the confluence of the Cuareim and Uruguay rivers and the Río de la Plata. The regulation of water uses, namely navigation, works, pilotage, top on and top off zones, bed and subsoil resources, fishing, pollution prevention, jurisdiction, and settlement of disputes procedures are expressly dealt with. The Administrative Commission set up under this agreement, Articles 49–57, to develop the river statute, was settled in Paysandú, Uruguay on 22 November 1978. The commission, known as CARU by its Spanish initials, developed several chapters of the statute. It also carried out water quality monitoring, with some 200,000 samples examined in its 18 years' functioning period. As there are always people behind the state's deliberations, it is not possible to think about these impressive agreements on the Plata basin in the decade between 1966 and 1975 without regard to the political "working group" behind the stage.

The dense chronology of facts enshrined in that period of constraint by the same state actors was not only the outcome of the Asunción Declaration but largely the influence of a group of highly qualified specialists, jurists, and diplomats of the five member states of the Plata basin, as well as those officials of international organizations, who arranged for the global conception of local needs and general principles to be drafted together, who understood the social and political pressures to give birth to new endeavours and who accepted the challenge to surmount existing opposition. Though not expressly quoted, not only power dams but necessary infrastructure works were built during the recent past in the Plata basin region, namely roads and badly needed national and binational bridges linking cities, people, and economies.

For those undertakings special binational commissions were established, among them the two bilateral commissions for the international bridges over the Uruguay river in the boundary stretch between Argentina and Uruguay, at the towns of Unzué Harbor (Argentina) and Fray Bentos (Uruguay), and Colón (Argentina) and Paysandú (Uruguay). That was also the case for the international bridge between Brazil and Argentina over the Iguazú river, at Foz de Iguazú and for the international bridge on the Paraná river between Paraguay and Argentina in the cities of Encarnación (Paraguay) and Posadas (Argentina).

The international commissions were dissolved after the works were completed, as they were public works not under concession, to be administered by each government's official departments. Therefore there was no longer any need for a permanent body for the control of and relationship with the private consortium.

Recently, other river commissions were agreed upon, referring to the international tributaries of the basin, the Bermejo and Pilcomayo rivers. The Pilcomayo river is a Bolivian river at its origin, a boundary river between Bolivia and Argentina up to the trinational Esmeralda point and a boundary river between Paraguay and Argentina in its lower stream, until the conflux with the Paraguay river. To manage the river efficiently the convergence of the three riparians is required. This is a decisive reason for the failure of previous bilateral efforts to overcome the difficulties of this body of water.

On 26 April 1994, it was agreed to constitute a Trinational Commission for the Development of the Pilcomayo River Basin. The agreement to inaugurate the commission was subscribed to by the riparians of the basin, Bolivia, Paraguay, and Argentina, on 9

February 1995, with an appointed Executive Director by consensus of member states. The commission is an international organization with its functions established in the above-mentioned agreement.

Almost coincidentally, on 14 September 1993, an Administrative Bilateral Commission was proposed for the boundary stretch of the Pilcomayo river between Argentina and Paraguay. The commission was agreed upon on 5 August 1994, with the main purpose of dealing with the projects and works necessary to bring to a standstill the process of siltation and subsequent overflow of the water body.

In reference to the Bermejo river, a Bolivian river at its source, a boundary river between Bolivia and Argentina in the next stretch and an Argentine river in its longest downstream stretch, there had been several agreements in the recent past between the two riparians, especially taking into account the upper basin of the river.

Most recently, on 9 June 1995, Argentina and Bolivia entered into an agreement for the development of the upper basin of the Bermejo river and its tributary the Río Grande de Tarija. The outcome of this agreement was the Binational Commission, an international body with management capacities for river works. The commission is entitled to draft, bid, finance, and administer the undertakings required for regional development.

On 19 November 1996, and in the framework of the precedent agreement of 9 June 1995, the Bermejo river basin riparians signed the complementary protocol for the construction of three under-takings. One of them is the dam, reservoir, and complementary works on a tributary of the Bermejo, the Tarija river, 17 km upstream at its confluence with the Itaú river, on the site called *Cambarí*. Another work is the Las Pavas complex on the upper Bermejo river in the boundary stretch between Argentina and Bolivia, 50 km upstream of the confluence with the Río Grande de Tarija river. The last is the Arrazayal dam, reservoir, and complementary works, about 25 km downstream from the Las Pavas undertaking, with the additional purpose of functioning as a counterbalance to Las Pavas.

The Plata basin treaty system

With reference to the institutional organization of the basin, we should first analyse the Plata Basin Treaty itself. The treaty has named two main bodies responsible for fulfilling the ambitious targets mentioned in its Article I. These organs have a political and diplo-matic level, one of them on a non permanent basis, the Conference of

Foreign Affairs Ministers, and the other of a permanent character and periodic meetings, the Intergovernmental Coordinating Committee. The Plata Basin Treaty, nonetheless, did not establish these bodies but granted them institutional permanence. Actually, the Plata basin countries' Foreign Affairs Ministers Conference had two meetings before gaining the treaty's approval, and the third one took place together with the special meeting for the treaty's signature. As regards the Intergovernmental Coordinating Committee (CIC), this had already been created by the Santa Cruz de la Sierra Conference in 1968 and further incorporated by the treaty, as it is "recognized as the permanent body for the basin" according to Article III.

The treaty additionally accepted, in Article III, paragraph 1, the statute for the CIC adopted at the Santa Cruz de la Sierra Conference, 1968, as the proper instrument to regulate the Committee's activities. As the statute proposed the Secretary of the Committee, this body became another organ of the treaty and completed the structure of the Plata Basin international organization. It has been also decided in the treaty that both the CIC and its Secretary will be located in Buenos Aires, Argentina.[45]

Thus, the Plata Basin Treaty does not establish new bodies but has organized the existing ones: the Foreign Affairs Ministers Conference, the Committee (CIC), and the Secretary. As regards the Foreign Affairs Ministers' Conference, this met 20 times, with four Extraordinary Conferences till 1996, and has approved 250 resolutions since the first conference in 1967.

The conference resolutions do not have mandatory character as regards member states, but they are applicable only in reference to the functioning of the Plata basin bodies. Among other things, they are compulsory when approving the budget, amending statutes, or incorporating new organs. This is similar to the majority of international organizations.

Invested as the permanent body of the organization inaugurated by the treaty, the committee has international capacity as to the fulfilment of its functions. This capacity is not expressly stated by Article III of the treaty or the statute itself, but it is recognized by the Settlement Agreement between the committee and the Argentine Government. By this means, both the international capacity to adopt agreements and the municipal legal capacity under Argentine law[46] are accepted.

The CIC has held 516 meetings, the last one on 11 December 1996, since its inauguration in 1968. This diplomatic body has maintained the continuity, though not the effectiveness, of the Plata Basin Treaty structure.

In 1992, the statute of the committee was amended[47] and another body, the Delegates Commission, was added. The Delegates Commission has a subsidiary relationship as regards the CIC and retains the diplomatic feature of the main body. As long as the committee sessions are attended by ambassadors and other senior diplomats, the commission meetings are attended by active members of the five member states' diplomatic staff.

The other modification was in reference to the Secretary, which has been transformed into the Executive Secretary. This amendment allowed the Secretary's administrative autonomy and incorporated budgetary as well as staff management functions. It would be desirable to add a technical office, but there was no agreement on this regard about the enlargement of the Secretary's functions.

As to the opportunities for drafting studies and works and obtaining financial assistance, the Plata Basin Treaty structure has had the most limited activity. At present, in 1996, a modest programme of a hydrological network is in the stage of receiving financial assistance from IDB, as is the case of another programme under consideration by FONPLATA.

As early as 1968 there were meetings of technical groups of the Plata basin states, and the Santa Cruz de la Sierra Act 1968, organized projects in A, B, and A/B categories in charge of specialized technical groups. The groups were later identified as Technical Counterparts and included in the restatement of the CIC Statute in 1992. The selected topics were water quality, flood alert, natural resources, namely soil and navigation. There was the requirement from Brazil, quite recently and not yet mandated, to incorporate the counterpart meeting for the environment. The projects at present under consideration by IDB and FONPLATA were drafted by the appropriate counterparts' meetings.

Because of financial constraints of member states these groups have not held any meetings since 1994. The budgetary incidence on the functioning of the system will ask for careful consideration in the future. For an international structure, the key to efficiency consists of functions to carry out, capacities to fulfil them, and a budget to perform those functions.

With a long-term vision the Plata Basin Treaty bodies[48] supported the creation of a special fund for the local financing of regional projects and the capacity to seek international financial assistance. The

Treaty for the Establishment of the Plata Basin Financial Development Fund, signed on 12 June 1974 and in force on 14 October 1976, set an authorized capital of US$1bn (Article 5), with US$20m to be integrated initially (Article 6). Brazil and Argentina will contribute 33 per cent each, and Bolivia, Paraguay, and Uruguay 11 per cent each. The remaining authorized capital will be disbursed by Argentina and Brazil in three years and by the other three members in ten years, in annual instalments. By a subsequent regulation, Fonplata approved a preferential system of credits to grant assistance to relatively less developed economies. From this policy, 84 per cent of the loans were given to Bolivia, Paraguay, and Uruguay. Carefully administering these scarce resources, Fonplata has granted loans for US$330bn of which it has paid US$167bn so far.

At present, during 1996, member states, headed by Argentina and Brazil, are negotiating a fundamental structural amendment of the Fonplata Constitutive Agreement. The IDB, which has recommended this transformation for several years, could provide the consultant support if member states asked for it. According to the resolution approved by the Foreign Affairs Ministers' Conference,[49] the pattern of substantive change should include: (a) the inclusion of other members, both from the basin or outside; (b) free subscription of capital and a vote capacity related to such capital; (c) financial capacity for the whole territory of Plata basin states, not only the Plata region; (d) a loan policy without restrictions, and (e) modern management and banking methods.

This foreseeable transformation will be positive for regional economies, though it will abandon the special financial system of the Plata basin for infrastructure works. There is no manifest intention of member states to leave aside the possibility of preferential loans for Plata basin projects. Nonetheless, a stronger financial institution would improve the elegibility of regional projects, either as basin infrastructure or as regional requirements not related to the basin itself.

The Paraguay-Paraná rivers are a natural transport corridor running north–south from the heart of South America to the Atlantic ocean, by the Plata river. The natural waterway was navigated from the sixteenth century onwards by the Spanish fleet, and the city of Asunción, on the Paraguay river banks, since 1537 was a centre for expedition departures in colonial times.

Only in 1988, encouraged by agricultural expansion in the Brazilian states of Mato Grosso, together with the mining development in

Brazil and Bolivia and economic growth in Paraguay, did interest appear in improving the conditions of the existing waterway, at the first meeting in Campo Grande, Brazil.

The river has been used insufficiently and inefficiently for many reasons, not only the watercourse conditions. The required improvements should imply ensuring a whole year's minimum depth, structuring difficult stretches and buoying the waterway with day and night signals.

The riparian states of the Paraguay and Paraná rivers are Brazil, Paraguay, and Argentina. Bolivia has a narrow stretch, connected by the Tamengo channel to the Paraguay river. Uruguay is not a riparian, and for that reason the New Palmira harbour,[50] on the Uruguay river bank, has been chosen as one of the extremes of the waterway. Consequently, the five member states of the Plata basin have started a new mechanism for this project, with the financial support of IDB and Fonplata.

To deal with this undertaking an Intergovernmental Waterway Committee (CIH) was established and became another body of the Plata basin structure in 1991. Until the present, the most important achievement of this committee was to produce the Waterway Transport Agreement, approved by member states in 1992 and entered into force on 13 February 1995. The Transport Agreement, with eight Protocols, is a common navigation code for waterway users applicable to five riparian states.

In December 1996, the feasibility engineering draft and the environmental impact assessment studies for the Paraguay-Paraná waterway was delivered to member states and future works are at the decision stage.

With the inclusion of the Paraguay-Paraná Waterway Committee the Plata basin institutional system incorporated the last link to its chain. Theoretically, all the bodies should be coordinated by CIC, but this is not the case. As far as Fonplata and the Waterway Committee are concerned, they are not only autonomous but almost independent of any tie with the CIC.

It would also be desirable that the already mentioned binational or trinational commissions should have an almost informal and flexible relationship with the Plata basin institutional framework, as all of them are related to the same water body. There is an important task to fulfil here, in order to bring together the pieces of this large puzzle.

The tragedy, called inefficacy, of international organizations is the contrast between extensive functions, not difficult to include in constitutive agreements, and extremely limited capacities. These include limited individual personalities, limited decision-making aptitude, limited staff, limited budget. No one would be surprised by limited results.

For the La Plata basin institutional system the present is a moment of challenge and change. If the desired changes are to take place, political decisions should be taken. It would be desirable to say that those decisions "shall" be taken, but there is no certainty about the compatibility of purpose of five different member states.

As a first step for the decision-making stage a document called the Restructuration of the CIC – Resolution No. 2 (IV-E) was prepared and approved by the CIC on member states. As a starting point it was admitted that it would be an oversimplification to consider that the integration process of the Mercosur renders the Plata basin system unnecessary and that its goals are coincident with those of the free trade zone. The document is divided into six chapters. The introduction defines the Plata Basin Treaty as a systematic and programmatic instrument (para. 5). It admits that the treaty does not include guidelines for water administration, but the institutional framework to allow the agreement on water administration patterns (para. 6). The second chapter describes the bodies created by the treaty and those added afterwards. Chapter III deals with the system's present situation and the achievements on waterworks and infrastructure, although there is a lack of cooperative bodies for water administration on the multilateral level (paras. 12 and 13). Furthermore, the CIC has not been updated to cope with new threats and demands, and it has gradually become weaker in its last stage (para. 15). Nonetheless, the Plata basin system has a role to play as regards building up a common policy for the basin states (para. 16). For that aim it is necessary to: (a) re-elaborate its goals; (b) modify its structure; and (c) design its funding (para. 17). Chapter IV defines the objectives of the system in reference to member states' interests and to the principles of the Dublin Conference on Water and Environment, convened in Dublin on 26–31 January 1992 and also addressed in the Morelia Declaration, adopted at the Water Basin Organization's meeting, held in Morelia, Mexico, on 27–29 March 1996. Those objectives are: (1) conservation and administration of water resources; (2) the region's physical integration; (3) data and information systematization; (4) environmental preservation; (5) harmonization of policies and legislations (para. 18). Chapter V proposes the future bodies of the Plata basin system. It maintains the Foreign Affairs Ministers Conference as the high level body and establishes the Plata Basin Commission as the substitute for the CIC, with a project unit which becomes an actual innovation substitute for the present non-efficient structure. The future Secretary General, envisaged as a position with an influential profile, will support both the newly created commission and the technical office (para. 19). Finally, chapter VI refers to the design of its funding, a veritable test for checking the political strength of every international organization. In addition to contributions from member states, it considers the financial assistance of hydroelectric power binational entities as well as the support for specific projects of international financial institutions (para. 21).

The document will be submitted to the Foreign Affairs Ministers Conference in 1997 for consideration and approval.

Water-resources planning

As planning in general is to try to get the most efficient use of available resources with the available means, water-resources planning demands an extremely careful consideration of both elements – resources and means. Besides, as water resources are vulnerable in so many senses, namely, quality, quantity, limited offer, complex governance, and increasing demand, caution is perhaps the best definition and the relevant attitude to adopt when approaching this task.

Though planning is usually applicable when considering economic growth, there is a substantial difference if this planning activity is biased towards developing economies. In a developed economy context planning is carried out for efficiency and growth, but in a developing one together with those features it also implies social and cultural change. This change is a precondition and a consequence of development. Under these circumstances, in developing economies and countries, planning is closely inserted in governmental activities which set their seal on the objective, complemented by private investment. As a consequence, public sector influence cannot be avoided and that stresses the point that for convenience at the governmental level, planning and decision making should be invested in the same body.

Another condition to be considered is the intrinsic relationship between land use and its impact on water resources. Nowadays, it is not possible to look at land use and water-resources planning separately. They have to be part of the same process at all levels.[51] This is another aspect where developed and developing economies have a different standpoint, since users do not have the same opportunities to afford the costs of adequate water management. A great amount of intelligence and common sense has to be applied to balance the competitiveness of the global economy with the long-range benefits of environmentally sound administration of natural resources. This aspect is especially critical for non-developed situations and the public sector's responsibility is in this regard unavoidable.

It should be remembered that as regards needs to be satisfied the most important issues for planning activities are labour supplies, energy, transportation, agriculture, cattle raising, industrial uses, mining, flood control, tourism, and recreation. The aim of an efficient planning activity is to arrange for these necessities with the least possible harmful social, economic, and environmental effects. This means policy-making, whose necessary tool has to be produced by a thorough collection of data, namely hydrological, meteorological, economic, and technical, duly processed and evaluated. At this point, the institutional organization of planning becomes significant, both in the national and the international field. In national administration, federal states have a particularly complex law-making structure which deserves special analysis. Nonetheless, the coordination of the different stages of regulation elaborated by national and international agencies is a meaningful assignment for water-resources policy making.

It is in this gap that a relevant contribution has to be made by those with the knowledge and the experience to advance suitable proposals.

Notes

1. UN Secretary General report, Doc. A3/36/356.
2. Took place at the Quitandinha Hotel, in Petropolis, Brazil, convened by OAS.
3. Carlos Sanz de Santamaría, 1971. *Revolución Silenciosa*, pp. 17–21. México: Fondo de Cultura Económica.
4. Ibid. pp. 215–219.
5. General Assembly Resolution 2029 (XX), 22 November 1965.
6. General Assembly Resolution 1995 (XIX), 30 December 1964.
7. Doc. OEA/ser.G/VI.C/INF.-231 (Spanish).
8. Doc. OEA/ser.I/VI.2,CIJ-79, in OEA Documentos Oficiales, OEA/Ser.I/VI.CIJ.75rev., pp. 140–156.
9. X Resolution, Río de Janeiro, 30 November 1965, ibid., n. 8, p. 5.
10. The Portuguese text says: "... os graves e profundos inconvenientes de fórmulas que procuram institucionalizar o que deve ser fruto de uma cooperaçao natural e espontanea, pois, até mesmo nos prazos em que cada país estima oportuno reagir

as sugestoes que lhe sao apresentadas, se exerce a plenitude de sua soberania," in *Aprovechamiento energético del río Paraná*, COMIP, Comisión Mixta Paraguayo Argentina del Río Paraná, Buenos Aires, 1992, p. 25.

11. IDB, *El Estudio Conjunto e Integral de la cuenca de la Plata*, Report to CIC, Buenos Aires, July 1968, 1–2/14–17.

12. CIC, Minutes No. 18 (17 January 1968), No. 22 (14 March 1968), No. 24 (31 May 1968), No. 40 (25 July 1968), No. 42 (1 August 1968), No. 45 (25 August 1968).

13. OAS, Secretary General, Legal Department. 1968. *Programa para el desarrollo de la cuenca del Plata: Aspectos Jurídicos e Institucionales*. December, p. 17.

14. Ibid., pp. 38–39.

15. UNTS. 1973. Vol. 875, pp. 11–13.

16. In this sense the Resolution No. 11-IV 1971, where governments and the CIC are encouraged to speed up their demands for technical and financial assistance to international organizations.

17. For example: Resolution No. 3-IV 1971 established a permanent technical group for navigation, with experts of the five countries, with the purpose of proposing measures and projects to facilitate navigation of the Paraguay, Paraná, Uruguay, and Plata rivers; Resolution No. 35-V 1972 convened an expert group in ichthyology, for the evaluation of fish resources; Resolutions No. 10-IV (1971) and No. 36-V (1972), installed an expert group on transportation; Resolutions No. 12-IV (1971), No. 39-V (1972), No. 40-V (1972), and No. 41-V (1972) settled an Ad-Hoc Technical Commission for Telecommunications.

18. Resolution No. 47-VI (1972) recommends a schedule of regular meetings for technical groups and commissions, with an established agenda and Resolution No. 48-VI (1972) asked the CIC to assess technical groups functioning in order to coordinate their activities.

19. Resolution No. 30-V (1972) on acknowledgement of Foreign Ministers Conference to OAS for the Final Report on *Cuenca del Río de la Plata – Estudio para su planificación y desarrollo – Inventario y análisis de la información básica sobre recursos naturales*.

20. Secretary General, OAS. 1969. Washington, D.C.

21. Secretary General, OAS. 1973. *República Federativa do Brasil, Noroeste do Estado do Paraná, Estudo para o Controle da Erosao*. Washington, D.C.

22. Secretary General, OAS. 1975. *República del Paraguay – Desarrollo de la Región Nororiental*. Washington, D.C.

23. Secretary General, OAS. 1974. *República Argentina-República de Bolivia, I. Alta Cuenca del Río Bermejo, Estudio de los Recursos Hídricos*. Washington, D.C.

24. Secretary General, OAS. 1977. *República Argentina, II. Cuenca Inferior del Río Bermejo. Programación para su Desarrollo*. Washington, D.C.

25. CIC, Minutes No. 96 (19 March 1970), 97 (2 April 1970), 98 (17 April 1970), and 99 (24 April 1970).

26. Jorge Nelson Gualco. 1972. *Cono Sur, elección de un destino*, p. 20. Buenos Aires: C.G. Fabril Editora.

27. UNTS. 1980. Vol. 875, p. 13. New York.

28. Bruno V. Ferrari Bono. 1996. "La Cuenca del Plata. Características. Su función integradora." In: *Documentos del Curso Desarrollo y Gestión de Cuencas Hidrográficas*, p. 293. Rome: Instituto Italo-Latino Americano (IILA).

29. Those were the Yacyretá-Apipé and the Corpus Christi dams.
30. Resolution No. 25-IV 1971. in OAS/Ser.1/VI-CIJ.75rev2, p.187.
31. Lake Lanoux Arbitration, *France* v. *Spain*, award of 16 November 1957, *International Law Reports*, vol. 24, pp. 101–138, at p. 134.
32. Frederic L. Kirgis, Jr. 1983. *Prior Consultation in International Law. A Study of State Practice*, pp. 85–87 and 359–362. Charlottesville: University Press of Virginia: Jon Martin Trolldalen. 1992. *International Environmental Conflict Resolution. The Role of the United Nations*, pp. 8–12 and 77–78. Geneva/New York: Unitar.
33. As is the case in Argentina, where the National Constitution, Article 26 (texts of 1853 and 1994) grounds freedom of navigation in navigable rivers according to national regulations.
34. Julio A. Barberis. 1971. "Principios jurídicos que regulan la libre navegación de la Cuenca del Plata," *Ríos y canales navegables internacionales*, pp. 197–209. Buenos Aires: Unitar.
35. Comisión Técnica Mixta de Salto Grande. 1994. *Documentos y Antecedentes, 1938–1994*, pp. 57–67 and pp. 31–41. Buenos Aires.
36. UN, doc. ST/LEG/SER.B/12, p. 163.
37. Comisión Mixta Paraguayo Argentina del Río Paraná, 1984. *Documentos Institucionales*, pp. 78–86. Buenos Aires.
38. 1 hectare = 10,000 sq. m.
39. UNTS. 1990. New York, vol. 1259, pp. 307–318.
40. About the Río de la Plata, cf. Edison González Lapeyre. 1997. *El Estatuto del Plata*. Fundación de Cultura Universitaria, Uruguay; Hector Gros Espiell. 1975. "Le Traité relatif au 'Rio de la Plata' et sa façade maritime." *Annuaire Français de Droit International*, vol. XXI, pp. 241–249; Lilian del Castillo Laborde. 1996. "Legal Regime at the Río de la Plata." *Natural Resources Journal*.
41. In this sense see bibliography referred to in note 39 above and also Ernesto J. Fitte. 1971. *Martín García, Historia de una isla argentina*. Buenos Aires/Barcelona: Emecé; and Antonio Emilio Castello. 1994. *Martín García, La gloriosa isla prisión*. Argentina: Universidad de Morón.
42. The length of the new channel is of 114 km from Km 0, Uruguay river down to Km 37, Plata river, with a depth of 32 ft and 100 metres breadth at the river bed. The selected Martín García channel runs between the Martín García island and the Uruguayan bank, and it is known as the Infierno (Hell) channel. There was another technically attractive option, to dredge the channel by the other side of the Martín García island, but it was not chosen.
43. Río de la Plata Treaty, Article 82.
44. On this subject, Frida M. Armas Pfirter. 1994. "El Derecho Internacional de Pesquerías y el Frente Marítimo del Río de la Plata." Consejo Argentino para las Relaciones Internacionales, CARI, Buenos Aires, pp. 331–362.
45. *Statute*, Secretary, Articles 6–9, Committee Headquarters, Articles 10 and 11.
46. *Settlement Agreement*, Argentine Government-CIC, Buenos Aires, Argentina, 22 May 1973.
47. XX Conference, Plata Basin Foreign Affairs Ministers Conference, Punta del Este, Uruguay, December 1992.
48. Resolution Nos. 5(IV)-1971 and 44(V)-1972 of the Foreign Affairs Ministers Conference.

49. Resolution No. 1-(IV-E), IV Extraordinary Conference of Foreign Affairs Ministers, 6 December 1995, Punta del Este, Uruguay.
50. Resolution No. 238, XX Conference of Foreign Affairs Ministers, Asunción, Paraguay, 1991.
51. Dante A. Caponera. 1992. *Principles of water law and administration*, pp. 161–168. Rotterdam: Balkema.

Abbreviations

IDB   Interamerican Development Bank
OAS   Organization of American States
UN   United Nations Organization
UNTS   United Nations Treaty Series

Annex

```
                        ┌─────────────────┐
                        │ Foreign Affairs │
                        │   Ministers'    │
                        │   Conference    │
                        └─────────────────┘

┌─────────────────┐    ┌─────────────────┐    ┌──────────────┐
│   Plata Basin   │    │ Intergovernmental│    │  Executive   │
│     Treaty      │    │   Coordinating   │    │  Secretary   │
│      1969       │    │    Committee     │    └──────────────┘
└─────────────────┘    └─────────────────┘

                        ┌─────────────────┐
                        │    Technical    │
                        │  Counterparts   │
                        └─────────────────┘

┌─────────────────┐
│    Fonplata     │
│ Plata Basin Financial│
│      Fund       │
│      1974       │
└─────────────────┘

┌─────────────────┐
│ Intergovernmental│
│    Waterway     │
│   Committee     │
│      1991       │
└─────────────────┘
```

Fig. 1   **Plata basin treaty system**

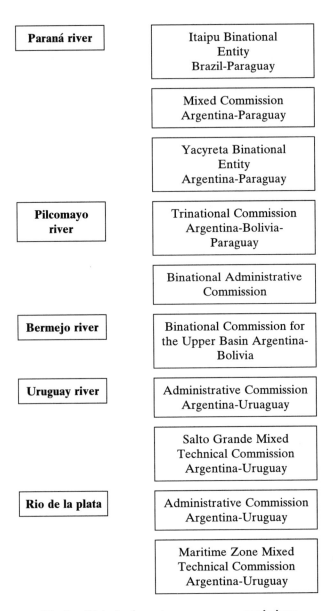

Fig. 2.   **Plata basin water-resources commissions**

Part 3
The São Francisco river basin

8

The Rio São Francisco: Lifeline of the north-east

Larry D. Simpson

Introduction

The Rio São Francisco rises in the cerrado of the state of Minas Gerais and Goias in the central region of Brazil fed by the runoff from orographic rainfall of the central plateau and chapadas that divide this drainage from the Tocantins and Amazon drainages to the North-west (map 1: National location map, valley of the São Francisco river[1]). The river arises in the state of Minas Gerais in the Serra da Canastra at an elevation of approximately 1600 m. From there it winds 2,700 km north and east through the semi-arid lands of the north-east region of Brazil crossing much of the area defined as the Drought Polygon of the country (map 2: Drought Polygon of the north-east of Brazil[2]) and provides a lifeline to the region. The flows of this drainage provide the hydropower to fuel the industry of the region, the water to supply the growing fruit and vegetable production industry, the transportation for goods and services, a fishery that provides fish renowned for their delicate flavour and fine texture, and the foundation for a culture unique in the world. The Afro-Brazilian spirits said to inhabit the river spawned the famous Kahunka figurines on the prows of the boats which plied the river during the last century in an effort to appease and ward off the more malevolent of

Map 1   **National location map, valley of the São Francisco river (CODEVASF)**

these spirits. The music of the people of this river system became a root for some of the delightful music of the region. This river system holds the key for the future of the region, but also represents one of the major potential sources of conflict as the many developing demands for scarce water supplies within the north-east of Brazil compete for the lifeblood of the river. These demands are not just limited to the riparian states of the river basin. The non-riparian semi-arid states of the north-east have long coveted the waters of this river system and proposals for major transbasin diversions to the north and east of the drainage have been put forth for over 75 years. The emotional, environmental, political, and economic struggles that such diversions proposals will spawn have just begun to emerge. This river system will be the subject of intense study, development, and controversy during the coming century and the solution of these controversies will require the best technical and political minds the

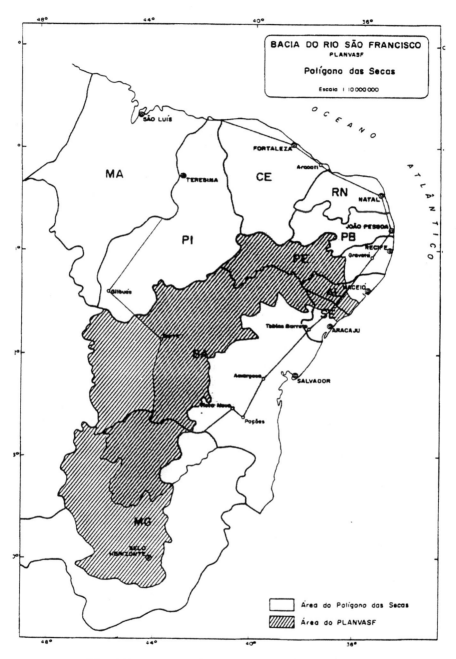

Map 2　**Drought Polygon of the north-east of Brazil**

209

country has to offer. The complexities of the system will require the use of the latest in computer and hydro-meteorological information technology to provide the decision makers and diplomats with the information and tools necessary to forge compromises and to develop and prioritize the competing and, frequently, conflicting uses of the resources of the basin. The challenge of meeting the multi-purpose demands for the water of the river system in a sustainable and environmentally acceptable manner will tax future thinkers and decision makers to the limit. The equally important challenge of providing for these demands in a manner that does not threaten or destroy the unique ecology of the river and pollute its scarce and valuable water supply will require compromise, sound planning, political and scientific cooperation, and a tremendous amount of effort, time, and financial resources.

Geography and climate

The basin of the Rio São Francisco encompasses a drainage of approximately 640,000 km^2,[3] with the drainage beginning in the high elevations of the Serra da Canastra in the southern part of the state of Minas Gerais and ending at its mouth to the Atlantic ocean as it forms the border between the States of Alagoas and Sergipe. The river is perennial in nature as are its primary tributaries in Minas Gerais and the western part of the State of Bahia. The river traverses climatological variations ranging from humid to arid. The main part of the flow of the river system is contributed from the humid and semi-humid drainages near its headwaters with only flood flows from the intermittent rivers of the arid and semi-arid regions adding to the flow during the rainy seasons. These contributions frequently provide more problems than benefits in the form of floods, erosion, and sedimentation in the river system. The river system has been subjected to a great deal of infrastructure modification, primarily for the production of hydroelectric energy, throughout its length. The basin has tremendous agricultural potential if its waters can be fully utilized for irrigation of the good soils along its margins. It also has the potential to provide a key transportation link through this region of Brazil and to provide additional energy to the region. These uses are not mutually exclusive but a great deal of coordination and prioritization will be required if the optimum mix of uses and management options are to be obtained from the asset values of the river system. The solution to this problem will involve the development of technological information and decision support tools and will also require resolution of problems of a more political and economic nature.

The Rio São Francisco basin is normally divided into four distinct sub-basins[4]: the alta or upper; the medio or middle; the sub-medio or lower middle; and the baixo or lower basin (map 3: Subdivision of basins, valley of the São Francisco river[5]). Each has distinct characteristics from a topographic and climatological standpoint.

Alta sub-basin

The alta or upper sub-basin is located in the most southerly part of the basin, primarily within the state of Minas Gerais and is a region of rolling hills, table lands, and high flat chapadas. The climate is temperate to sub-tropical and can be classified as a humid zone. The precipitation in this part of the sub-basin averages approximately 1,250 mm per year with highs in the western part of the sub-basin of 1,400 mm per year and lows near the northern edge of 1,000 mm per year. The temperature within this sub-basin averages between an average low of 18 °C during the winter months to an average high of 23 °C during the summer months. The capital of the State of Minas Gerais, Belo Horizonte is located in this area with its strong industrial and commercial centre and the attendant production of wastes which eventually find a way into the waters of the river system. Petroleum refining is an important economic factor in both Belo Horizonte and in the city of Betim located in the southerly extreme of the sub-basin. Mining for construction materials also predominates within this region. The average altitude is between 1,000 to 1,300 m above sea level with high flat chapadas along the western edge and rolling hills to the centre and south-eastern part. The vegetation is typically cerrado or savanna with good soil and rich agricultural production potential. Where irrigation has been developed, a strong fruit culture industry is developing. This is particularly prevalent in the northern edge of the sub-basin in the vicinity of Janauba and Gorituba where strong private irrigation has been coupled with well-developed federal efforts to develop a high production fruit growing area that exports internationally and to the more highly populated areas of Brazil. Production from the high chapada areas to the west is primarily soya and cattle with some gradual shift to higher value crops within irrigated areas. This region has also produced a large amount

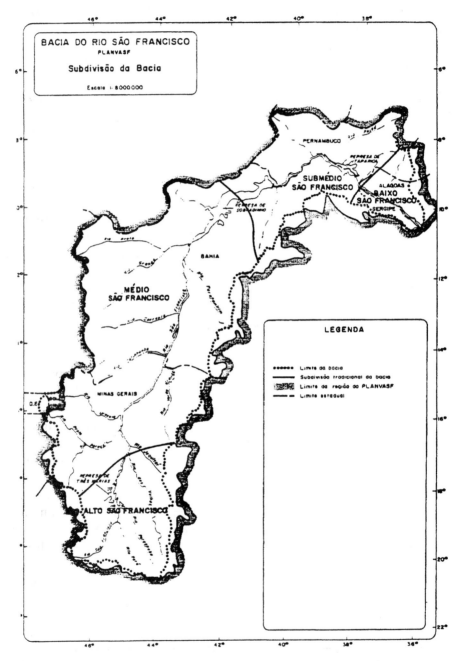

Map 3   Subdivision of basins, valley of the São Francisco river

of cultivated forests of eucalyptus for use in the paper industry and in the production of charcoal for the steel industry. The rivers within this sub-basin are predominately perennial in nature and are the primary sources of flow for the rest of the river basin. Infrastructure development within this reach of the river includes irrigation diversion and the large Tres Marias dam and reservoir. This facility is used primarily for hydroelectric energy production and for the regularization of the river flow to control flooding in the river system below. This reach of the river contains the basin's largest urban population located in the metropolitan Belo Horizonte area. While other cities are also located within this upper drainage, the impact of Belo Horizonte vastly overshadows all other factors in terms of existing and potential pollution, economic influence and population. Other moderate-sized cities located within this sub-basin include Patos da Minas, Januaria, Betim, and others. With a population of almost 7 million based upon a 1994 census[6] this region represents over half of the total population of the São Francisco river basin.

Medio sub-basin

The medio sub-basin of the river is located within the States of Minas Gerais and Bahia and is characterized by two distinct zones. On the left or western margin of the basin, the western zone is fed by orographic rainfall in the high areas to the west. The tributaries are primarily perennial with the major river systems being the Corrente river and the Grande river. The vegetation is primarily cerrado in the upper and left margins of the sub-basin and caatinga or semi-arid type vegetation in the remainder of the sub-basin. The soils in this area are excellent for agricultural production and both private and public irrigation has developed historically. The left margin in the southern part of this sub-basin is strongly influenced by the perennial tributary river systems with significant irrigation development on these tributaries in both the States of Minas Gerais and Bahia. However, the right or eastern margin of the sub-basin within the State of Bahia is characterized by intermittent or seasonal tributaries and has considerably less development. It is predominately caatinga vegetation characteristic of the sertão or semi-arid area. Primary use of the land in this area is for cattle and goat production with limited subsistence agriculture and irrigated agriculture where water is available from surface or underground sources. The Rio São Francisco through this sub-basin is not broken by any dams or reservoirs until its lower or northern extreme. At the lower reach of the sub-basin, the head-waters of the Sobradinho reservoir are encountered. The land and soils of this sub-basin represent a significant potential for the development of irrigated agriculture in the north-east region of Brazil. The large amount of solar radiation, moderate to hot temperatures, and good soils provide a high potential for the development of high value fruit, vegetable, and coffee crops as well as major soya and grain production. The development of this potential could provide a strong economic boost to the region and to Brazil as well as providing a major source of food for less productive areas of the world. The precipitation within this sub-basin averages around 900 mm per year and the average temperature ranges from 25 °C in the summer to 23 °C in the winter. The population is rural and sparse by comparison to the upper basin. The population of the area is principally involved in agricultural-related activities and dependence on the river for irrigation, transportation, and water supply predominates. Without the influence of the larger metropolitan areas on the income statistics, the average income of this sub-basin is much lower than that of the upper basin and it is estimated that over 50 per cent of the families within this sub-basin are within the category of indigent poor, based upon a study done by the IPEA in 1993.[7] The key to the resolution of this problem is improved education, vocational training and the creation of employment and the development of economic opportunities within the region. This will require the wise and sustainable management of the water supplies of the region to support any economic advances within the sub-basin.

Sub-medio basin

This sub-basin is located in the middle to lower reaches of the basin in the States of Bahia and Pernambuco. The river in this sub-basin forms the border between the two states and represents a major source of irrigation for the famous fruit and vegetable production region of Petrolina/Juazeiro. The major federal and private irrigation projects located in the Petrolina/Juazeiro area along with the strong associated agro-industry represents one of the success stories of irrigated agriculture within Brazil. Fruit and vegetable crops are exported from this area throughout Brazil and to foreign countries. The development of the region was strongly influenced initially by the development of federally sponsored irrigation projects which provided the base for subsequent private investment. The federal effort was primarily the responsibility of the Companhia de Desenvolvimento do Vale do São Francisco (CODEVASF), a federal public company.[8] While much has been said of the inefficiency and waste involving federal involvement in large irrigation projects, it is doubtful that this highly successful private enterprise zone in agro-industry would have developed without the seed of federal investment. The emancipation of these federal efforts now represents a potential for the continued expansion of this success. The present efforts of CODEVASF in the middle basin in the State of Bahia near the cities of Barreiras and Bom Jesus da Lapa hold the same potential for providing the seed for major private investment and development. The creation of employment within these areas should have as significant role in the reduction of poverty as it has had in the Juazeiro/Petrolina region. It has been estimated that approximately five jobs have been created for each hectare of land that has been put under irrigation within this region. This is directly attributable to the strong development of high-value irrigated agricultural production within the region along with the associated food-processing industry with improved transportation and marketing.

This sub-basin also contains the majority of the development for hydroelectric energy production within the São Francisco basin.[9] The major dams and reservoirs of Sobradinho, Itaparica, Paulo Afonso, and Xingo provide renewable hydroelectric energy for most of the north-east of Brazil. The lower reach of the river within this sub-basin has a significant gradient and, therefore, has provided a strong potential for the development of hydroelectric energy. While this energy does not completely satisfy the demand within the São Francisco river basin, it provides the majority of the energy now needed and represents a stable foundation of renewable electric energy for the future.

The river within this sub-basin was originally marginally navigable through its upper reaches. With the existence of these storage and hydroelectric facilities, the regularization of flows to accommodate navigation represents a potential, if adequate coordination and management can be arranged for the multi-purpose use of the water.[10] The reaches primarily affected by the hydroelectric development, with the exception of Sobradinho reservoir, were not originally navigable as the gradient and rapids precluded commercial navigation.

It is within this sub-basin, at a point below Sobradinho reservoir, that a transbasin diversion project has been proposed to supply water to the north-east states of Ceará, Rio Grande do Norte and Paraíba. This project has been proposed in several forms for many years but has consistently met with political opposition as well as economic and financing obstacles. A scaled-down version of this proposal was presented within the last two years that presents a more technically and economically reasonable project. However, this proposal has still not resolved the problem of opposition from the riparian states within the basin. Recent studies by a commission of the federal legislature re-examined this proposal as a part of the major study of the overall development potential for the entire river basin.

Within this sub-basin, all of the tributary rivers from both the left and right margins of the river are seasonal in nature and their flow contribution to the overall basin flow is minimal.

The principal municipal concentrations of population are located in the cities of Juazeiro in Bahia and Petrolina and Paulo Afonso in Pernambuco. The remainder of the communities and population within this sub-basin are mostly rural and dependent on agriculture and livestock production.

The vegetation of this semi-arid area is predominately caatinga and the soils are predominately thin and non-productive. Areas such as the Petrolina/Juazeiro agricultural region are an exception to this but the majority of the sub-basin is mostly suited to limited livestock production. Mineral production from this region of the São Francisco basin has also been historically limited and probably does not represent a significant economic factor within this sub-basin. Precipitation within the basin averages about 500 mm per year and the temperature averages about 27 °C during the summer to 24 °C during the winter period. The stability of the temperature along with the almost continuous solar radiation provides a good climate for agricultural production as long as good soil and water are available and are properly managed.

Baixo sub-basin

The baixo or lower sub-basin of the Rio São Francisco includes the States of Bahia, Alagoas, Sergipe, and Pernambuco. The river forms the border between the states of Sergipe and Alagoas through this sub-basin. The vegetation of the lower basin is primarily cerrado and mata atlantica with transitional zones between the cerrado and the caatinga of the upstream sertão areas. The river in this lower sub-basin was historically navigable and was utilized for the transportation of sugar cane production, other agricultural production, and limestone and building materials. Since the upstream control of flows for hydroelectric production and the gradual decline of the Federal Navigation Company of the São Francisco (FRANAVE), the use of the river for commercial navigation has gradually diminished. The lower reaches of this sub-basin contain significant coastal deltas and coastal wetlands. Some of this area has been developed for agricultural use through the use of polder construction and drainage systems. The lower sub-basin is humid in its lower reaches in the mata atlantica but contains significant semi-arid drainages on its upper left margin with principally caatinga vegetation. Precipitation in this reach varies from 1,300 mm along the coast to 500 mm along the upstream boundaries of the sub-basin. The sub-basin population is principally concentrated near the coast in smaller municipalities and in rural communities. The average income of the population within this reach is low with the majority of the population falling in the category of poor and indigent.[11] The lower reaches of the river system have historically been used for transportation of agricultural products to markets along the coast, and the fisheries represent an important resource for this region. The ecological regime within the delta and coastal margins of the river also represent an asset that has not been fully defined or protected. The beach regions to the south of the delta represent a major nesting area for threatened and endangered sea turtle species.

Indigenous populations

The São Francisco river basin was home to an estimated 26,000 indigenous inhabitants, according to a survey done in 1988.[12] This represents approximately 0.6 per cent of the rural population of the basin. Of this indigenous population, approximately 19,495 or 74 per cent lived in the downstream reaches of the middle sub-basin and the lower basin in the States of Pernambuco, Alagoas, and Sergipe. These are predominately rural agrarian populations that are dependent on the land for their livelihood. The majority are involved in subsistence livestock raising and dry farming or in woodcutting with some crafts production. A significant population within Pernambuco was severely impacted by the construction of hydroelectric production systems such as the Itaparica Project and adequate mitigation and amelioration of these impacts has yet to be accomplished. The indigenous populations of the remainder of the basin represent less than 0.2 per cent of the total rural population of the basin. The economic and social condition of these indigenous populations is in reality not a great deal different than that of the majority of poor rural inhabitants of the basin.

Hydrology of the river system

The annual discharge of the Rio São Francisco at its mouth averages over 94,000,000 mil m^3 per year (See tables 1 and 2, Flows of Rio São Francisco and its principal tributaries[13]). The natural flow in the reaches through the middle basin below the principal perennial tributaries average between 2,100 m^3/second and 2,800 m^3/second and with a natural flow of approximately 3,000 m^3/second near the mouth of the river in the lower sub-basin. Normal natural maximum flows occur during the month of March and average approximately 12,950 m^3/second at Juazeiro near the boundary between the middle and sub-middle sub-basins and 12,967 m^3/second at Pão da Acucar located near the mouth of the river. Normal minimum flows at these stations occur during the month of September and average 671 m^3/second and 842 m^3/second at these two locations, respectively.[14] As the river is operated today, the natural flows are highly regulated by the extensive hydroelectric developments and are regulated to optimize energy production and to control flooding of the river margins.

In addition to the mainstream of the river, flows in the upper sub-basin are contributed from the major tributaries of the Das Velhas river (average flow 292 m^3/second), the Paracatu river (average flow 436 m^3/second) and the Urucuia river (average flow 251 m^3/second). The upper sub-basin contributes over 70 per cent of the overall flow of the river basin. The average annual flow of the river at the lower boundary of this sub-basin at the Manga gauging station is 2,050 m^3/second (table 1: Flows in the Rio São Francisco).

Within the middle sub-basin, two principal tributaries contribute substantially to the flow of the river. These are the Corrente river with average flows of 251 m^3/second and the Grande river with average flows of 262 m^3/second. With the exception of the Das Velhas river in the upper sub-basin, all of the principal tributaries flow into the river from its left or western margin where orographic precipitation forms the majority of the water source (table 2: Flows in the principal tributaries of the Rio São Francisco).

Table 1 Valley of the São Francisco, characteristics of the principal rivers

River	San Francisco bankside	Station	Period of registration	Distance from source to ocean[1]	Drainage area	Average annual discharge	Specific average discharge
São Francisco	–	Trés Marias	38–82	2.221	49.750	707	14.21
Das Velhas	Right: Direita	Pirapora	38–81	2.050	61.880	768	12.41
São Francisco	–	Várzea da Palma	38–75	2.025	25.940	292	11.25
São Francisco		Barra do Jequitaí	63–78		90.990	1.015	11.15
Jequitaí	Direita	Jequitaí	67–75	2.001	6.811	46	6.75
São Francisco	–	Cach. da Manteiga	59–81		107.070	1.132	10.57
Paracatu	Left: Esquerda	Porto Alegre	52–75	1.926	41.709	436	10.45
São Francisco	–	São Romão	53–81	1.893	153.702	1.520	9.89
Urucuia	Esquerda	Barra do Escuro	55–75	1.866	24.658	251	10.18
São Francisco	–	São Francisco	43–81	1.830	182.537	2.082	10.40
	–	Januaria	34–70	1.750	191.700	2.168	11.31
		Manga	32–81	1.645	200.789	2.050	10.21
Verde Grande	Direita	Boca da Caatinga	72–75	1.525	30.174	19	0.62
Carinhanha	Esquerda	Juvenília	64–78	1.586	15.832	150	9.47
São Francisco	–	Carinhanha	27–81	1.586	251.209	2.207	8.79
Corrente	Esquerda	Pt.° Novo	77–84	1.447	31.120	251	8.07
São Francisco	–	Morpará	47–79	1.234	344.800	2.421	7.02
Grande	Esquerda	Boqueirão	33–79	1.178	(65.900)	262	3.98
São Francisco	–	Barra	25–77	1.178	421.400	2.652	6.29
		Juazeiro	29–79	759	510.800	2.731	5.35
		Pão de Açucar	26–79		608.900	3.001	4.93
	–	Traipu	38–79	171	622.600	2.980	4.79

[1] Or from the station when located on the river San Francisco.
PLANVASF.

Table 2 **Valley of the São Francisco, average monthly flows**

River	Station	Registration period		Jan.	Feb.	Mar.	Apr.	May	June	July	Aug.	Sept.	Oct.	Nov.	Dec.	Year	Drainage area km²	Months without data
São Francisco	Pirapora	38–81	Med.	1,423	1,318	1,168	886	594	512	440	389	382	464	736	1,165	768	61,880	38
			Max.	4,006	3,134	3,096	2,068	1,745	1,515	857	920	786	971	1,782	2,062	1,647		38
			Min.	430	545	440	321	202	154	151	142	141	142	208	467	415		
São Francisco	Barra do Jequitai [Montante]	63–78	Med.	1,909	1,580	1,250	945	655	630	626	611	631	844	1,241	1,414	1,015	90,990	04
			Max.	3,919	2,842	3,584	1,578	1,017	928	869	920	857	1,133	2,261	2,017	1,483		
			Min.	687	830	494	394	373	296	288	292	255	470	534	557	683		
São Francisco	Cachoera da Manteira	59–81	Med.	2,225	1,849	1,647	1,130	773	705	645	603	598	792	1,310	1,419	1,132	107,070	16
			Max.	4,419	3,384	3,928	1,949	1,256	1,301	984	946	996	1,181	2,854	4,211	1,574		
			Min.	871	919	607	462	380	306	264	223	198	282	406	576	755		
Sao Francisco	São Romão	53–81	Med.	3,022	2,618	2,307	1,642	1,022	882	812	712	680	980	1,638	2,385	1,520	153,702	38
			Max.	6,024	6,887	4,873	4,049	1,846	1,642	1,396	1,250	1,170	1,572	4,015	4,842	2,614		
			Min.	1,017	1,127	896	613	459	380	295	233	186	301	501	699	944		
São Francisco	São Francisco	43–81	Med.	3,995	3,417	3,366	2,442	1,409	1,139	960	812	750	954	1,976	3,208	2,082	182,637	71
			Max.	9,689	8,446	7,484	7,814	2,706	1,936	1,510	1,314	1,373	1,586	4,794	7,870	3,492		
			Min.	1,104	1,269	1,098	882	701	557	438	369	300	331	603	972	1,344		
São Francisco	Januaria	34–70	Med.	4,321	3,985	3,574	2,614	1,553	1,164	981	797	703	955	1,921	3,459	2,168	191,700	03
			Max.	8,002	8,817	7,912	7,070	3,522	2,349	1,903	1,520	1,273	2,006	3,492	5,954	3,781		
			Min.	1,791	1,463	1,420	1,101	614	498	417	324	260	262	493	880	1,230		
São Francisco	Pedra de Maria da Crut	72–81	Med.	3,723	3,306	2,810	2,531	1,523	1,366	1,179	1,053	1,000	1,377	2,423	2,791	1,981	191,063	11
			Max.	6,924	7,814	5,524	3,413	2,154	2,077	1,560	1,403	1,442	1,794	4,454	3,573	2,937		
			Min.	1,222	1,434	1,202	1,013	1,083	903	879	733	707	1,088	1,320	1,541	1,454		
São Francisco	Manga	32–81	Med.	4,087	3,678	3,283	2,554	1,428	1,073	880	722	674	902	1,837	3,259	2,050	200,789	84
			Max.	8,820	10,024	6,913	7,781	3,471	2,048	1,586	1,230	1,128	1,659	4,262	6,226	3,737		
			Min.	1,180	1,335	1,210	844	712	573	425	346	332	303	481	1,037	1,188		

Location	Period	Stat.														Total	N
São Francisco Carinhanha	27–81	Med	4,286	4,179	3,630	2,803	1,640	1,228	1,040	683	797	904	1,900	3,421	2,207	251,209	22
		Max.	8,883	10,207	7,532	8,163	4,347	2,312	1,807	559	1,544	1,754	3,786	6,827	4,079		
		Min.	1,241	1,405	1,274	910	691	598	508	413	379	385	607	1,120	1,246		
São Francisco Bom Jesus da Lapa [Paratinga]	77–84	Med	4,888	5,814	4,549	3,742	2,149	1,794	1,521	1,342	1,285	1,552	2,293	3,611	2,878	273,750	–
		Max.	6,351	10,045	8,196	5,421	3,679	2,331	1,996	1,659	1,544	1,943	4,019	6,205	4,035		
		Min.	2,947	2,130	1,382	1,540	1,265	1,104	957	789	777	1,290	1,480	1,431	1,688		
São Francisco Morpará	45–79	Med	4,252	4,510	4,198	3,206	1,826	1,422	1,229	1,082	1,007	1,161	1,917	3,324	2,421	344,800	10
		Max.	9,068	11,207	12,327	7,578	3,675	2,454	1,929	1,722	1,741	1,875	3,486	5,476	4,328		
		Min.	1,710	1,609	1,536	1,131	904	783	702	617	555	532	749	1,249	1,513		
São Francisco Barra	25–77	Med	4,737	4,781	4,423	3,709	2,286	1,641	1,404	1,243	1,137	1,271	2,052	3,630	2,652	421,400	19
		Max.	8,690	8,949	10,747	12,717	8,425	4,156	3,079	2,556	2,165	1,926	3,669	6,128	4,655		
		Min.	1,694	1,811	1,825	1,366	1,152	975	817	740	680	696	865	1,477	1,726		
São Francisco Juazeiro	29–79	Med	4,462	4,874	4,708	3,937	2,510	1,720	1,461	1,292	1,165	1,265	1,971	3,413	2,731	510,800	09
		Max.	7,843	9,981	12,950	8,840	8,744	3,937	2,589	2,129	2,436	2,393	3,518	5,590	4,798		
		Min.	1,527	1,505	1,356	1,534	1,235	1,018	895	793	671	639	877	1,349	1,694		
São Francisco Petrolindia	77–79	Med	3,133	4,009	5,762	4,397	2,315	1,816	1,581	1,595	1,727	1,869	2,341	2,269	2,692	586,700	–
		Max.	4,329	6,939	12,364	9,549	2,701	2,207	1,809	1,803	2,071	2,059	3,043	2,608	4,290		
		Min.	1,480	1,599	1,650	1,761	1,860	1,526	1,447	1,441	1,522	1,752	1,647	1,599	1,835		
São Francisco Pão da Açucar	26–79	Med	4,714	5,290	5,371	4,632	3,038	2,018	1,680	1,475	1,347	1,377	1,944	3,407	3,001	608,900	11
		Max.	8,060	11,502	12,967	9,371	9,865	5,039	3,023	2,442	2,457	2,667	3,320	5,723	5,303		
		Min.	1,422	1,638	1,772	1,764	1,419	1,073	924	842	760	725	899	1,461	1,721		
São Francisco Traipu	38–79	Med	4,534	5,224	5,400	4,646	2,941	1,990	1,662	1,461	1,313	1,339	1,882	3,311	2,980	622,600	07
		Max.	7,825	12,152	13,743	9,384	10,205	5,101	2,901	2,308	1,927	1,964	3,382	5,529	5,244		
		Min.	1,487	1,705	1,705	1,750	1,501	1,108	941	804	690	644	795	1,333	1,768		
São Francisco Iborirama	77–80	Med	4,557	7,568	6,004	3,347	2,157	1,898	1,598	1,388	1,319	1,619	2,116	3,200	2,951	322,600	03
		Max.	5,613	11,914	11,589	4,965	2,673	2,369	1,803	1,587	1,618	1,778	2,889	4,073	4,230		
		Min.	3,070	3,763	1,600	1,652	1,487	1,271	1,125	936	924	1,462	1,635	2,563	1,674		
São Francisco Tres Marias	38–62	Med	1,508	1,417	1,194	776	463	350	287	231	216	296	616	1,153	707	49,750	01
		Max.	3,245	3,857	2,952	1,499	818	581	455	353	369	587	1,518	2,291	1,147		
		Min.	259	200	331	202	130	158	116	93	108	234	137	320			

Table 2 **(cont.)**

River	Station	Registration period		Jan.	Feb.	Mar.	Apr.	May	June	July	Aug.	Sept.	Oct.	Nov.	Dec.	Year	Drainage area km²	Months without data
Das Velhas	Varzea da Palma	38–75	Med.	631	486	437	265	172	133	112	93	89	146	318	600	292	25,940	13
			Max.	1,756	973	1,118	646	394	262	215	181	157	301	664	1,411	544		
			Min.	159	115	131	101	61	71	52	43	51	57	89	122	142		
Jequitai	Jequitai	67–75	Med.	89	71	69	32	14	10	9	8	7	31	91	119	46	6,811	
			Max.	181	153	169	46	23	18	16	14	14	61	201	255	57		01
			Min.	29	13	21	12	5	4	5	5	4	11	30	42	37		
Paracatu	Porto Alegre	52–75	Med.	921	849	774	485	284	213	167	134	109	178	406	691	436	41,709	
			Max.	2,050	1,856	2,070	1,408	623	433	342	269	233	309	801	1,727	857		30
			Min.	186	167	220	147	83	81	62	55	61	68	127	160	267		
Urucuia	Barra do Escuro	55–75	Med.	460	439	448	306	150	99	80	63	53	128	315	490	251	24,658	
			Max.	927	969	849	559	278	157	122	109	99	277	554	826	325		19
			Min.	137	125	129	122	54	51	43	35	37	31	96	185	169		
Verde Grande	Boca da Caatinga	72–75	Med.	57	24	19	30	9	5	4	3	3	7	24	43	19	30,474	
			Max.	77	36	37	52	10	7	5	5	4	8	31	63	25		
			Min.	40	14	6	17	7	4	3	3	2	4	18	16	14		–
Carinhanha	Juvenilia	64–78	Med.	178	181	181	163	134	123	118	112	109	128	172	205	150	15,832	
			Max.	278	284	293	214	159	151	136	130	127	155	239	286	181		03
			Min.	122	128	112	109	109	102	100	88	97	109	126	151	125		
Corrente	Porto Novo	77–84	Med.	312	342	295	291	233	218	206	198	193	213	245	277	251	31,120	
			Max.	375	549	344	356	256	250	230	220	208	242	321	340	291		–
			Min.	228	225	169	190	184	166	157	152	150	176	183	207	187		
Grande	Boqueirao	33–79	Med.	325	339	336	310	254	220	209	200	193	208	251	307	262	65,900	
			Max.	571	671	604	572	376	305	281	267	262	266	341	464	368		08
			Min.	235	242	224	214	188	168	165	160	165	171	181	229	128		
Pajeu	Jazigo	64–75	Med.	3	13	48	74	35	8	4	2	0.6	0.1	0.3	0.2	15	6,170	
			Max.	15	56	151	348	148	25	13	6	3	0.7	2.0	1	53		06
			Min.	0	0.4	1	0.3	0.2	0	0	0	0	0	0	0	1.0		

DNAEE (National Department of Water and Electric Energy Departamento Nacional de Aguas e Energia Eletrica).

Institutional and political situation

As was previously mentioned, the Rio São Francisco flows through five different states of the north-east. In actuality, it can also include the Federal District of Brasília as one of the headwater tributaries begins within the Distrito Federal. The states have varying degrees of institutional development in the area of water resources, with the States of Bahia and Minas Gerais being most advanced.

The State of Minas Gerais includes an agency with the responsibility of planning and management of water resources. This agency, the Department of Water Resources, is a part of the Secretariat of Minerals, Water and Energy. The state has begun a cooperative process of preparing a master plan for the management of water resources. It is also in the process of formulating a water law that will govern the issuance of water rights and the administration and management of the water resources. Various other entities including Rural Minas have been active in promoting the development and use of water within the state.

The State of Bahia passed a comprehensive water law in 1995 that created a Superintendencia of Water Resources within the Secretariat of Water Resources. This law clearly spells out the policy of the state in water resources and provides for the issuance of water rights, permits for infrastructure construction, dam safety, and water tariffs. This law is one of the most comprehensive in Brazil. The state has also completed master plans or "Planos Diretores" for the majority of the significant river basins within the state and is embarking on the development of a water-resource management project for those basins considered to be of highest priority within the state.

The State of Pernambuco recently formed a Directorate of Water Resources within the Secretariat of Science, Technology and Environment. This agency is beginning the preparation of water-resources master plans for the state as well as working on a comprehensive water law. It has progressed extremely well in the formulation of an extensive computerized water-resource database system that includes both surficial and underground water-resources data.

The States of Alagoas and Sergipe are in the process of delineating water-resources responsibility within the state and are beginning work on the planning and institutional structure to provide good water management.

The principal federal entities with responsibility within the basin are CEEIVASF, the Executive Committee for Integrated Studies of the Hydrographic Basin of the Rio São Francisco; CODEVASF, Companhia de Desenvolvimento do Rio São Francisco; CHESF, Companhia Hidroelétrica do São Francisco, the major power agency of the basin and SUDENE, an organization formed for the purpose of comprehensive planning in the north-east.

CODEVASF was created for the purpose of developing the economy of the São Francisco basin. Historically it has principally been involved in the development of irrigated agriculture and agribusiness within the region. This has included some major successes such as the Nilho Coelho Project in the Petrolina/Juazeiro area with its successful high value crops. It has also included some smaller projects of less success where the efficiency of the irrigation systems is questionable and the cropping pattern continues to be low-value crops and subsistence farming. This has resulted in areas that are producing at a level of value far below that which would justify the extremely expensive federal investments in the projects. In recent years, there has been a movement within the CODEVASF projects toward greater participation by user-governed water districts that have accepted responsibility for the operation and maintenance of the systems and the responsibility to set and collect water-user charges sufficient to maintain these projects at a sustainable level. This move toward decentralization has improved the level of maintenance and the rate of collection of the water charges. It has also resulted in a greater consciousness on the part of the users of the real cost of operation and maintenance of these systems. As a result, the sustainability of the systems has increased. CODEVASF has, for many years, concentrated on being a development and operations agency. However, an analysis by a special committee appointed by the legislature to review the future of the São Francisco basin has strongly recommended that this role should shift to one of promotion and coordination with greater empowerment of the user organizations and the private sectors.[15] This shift will add a greater impetus to the development of the role of the private sector in this basin. CODEVASF has a strong regional presence within each of the sub-basins and can represent a strong focus-point for the development of all sectors of the economy of the basin. CODEVASF has made some excellent progress in the development of the use of Geographic Information Systems (GIS) in the analysis of soils, cropping patterns, river morphology, and land use within the irrigated areas of the basin. This system could provide a foundation for the analysis of the entire basin from the standpoint of

potential development, environmental protection, watershed reclamation, water pollution control, and navigation. While a great deal
of additional work would be required to add these functions to the
database, the increase in reliability of planning would pay off in the
long run. CODEVASF has also been responsible for a number of
successful innovations in the area of training and assistance to the
small farmers and the youth of the project areas. This has included
a youth vocational training centre in the Formoso A and H projects
near Bom Jesus da Lapa in Bahia and a very successful revolving
fund investment programme to assist small farmers in beginning to
produce high value fruit crops and to learn the technology for fruit
production and new technology for localized irrigation systems.[16]
While this agency has its successes and its problems, the overall result
of its activities within the São Francisco basin has been very positive.

In 1959, the Superintendencia para o Desenvolvimento do Nordeste (SUDENE) was created and given a wide range of responsibilities, including coordination of all ongoing activities and investments
in the region, including the Drought Polygon of the São Francisco
basin. These efforts culminated in the mid-1970s with the creation of
several new programmes and financial mechanisms aimed primarily
at settlement, land distribution, and agro-industrial modernization.
This trend continued through the 1980s as the Federal Government
created several additional programmes for the rural north-east.
SUDENE's influence in the comprehensive planning for the basin has
diminished with the growing strength of the planning and development capabilities of the states of the north-east.

Hydroelectric energy from the Rio São Francisco

CHESF is the agency that is responsible for the development, operation and maintenance of hydroelectric generation and the bulk
energy distribution throughout the north-east. CHESF operates
plants with approximately 7,800 MW of installed capacity in the Rio
São Francisco basin and its tributaries with an additional 2,500 MW
under construction and a planned total future capacity of over 26,000
MW.[17] CHESF works in close cooperation with the state electrical
companies in each state and in some instances, has transferred generation responsibility to the states, i.e. Tres Marias power plant,
transferred to CEMIG, the State Electrical Company of Minas Gerais.
As was previously explained, most of the hydroelectric generation in

the São Francisco river basin is located in the sub-middles sub-basin. The hydraulic gradient in this reach is sufficient for the installation of efficient hydroelectric facilities and the flow in the river at this point provides a stable source of energy. The installation of this hydro-electric system has not been without some problems, however. The Itaparica Hydroelectric Project, located in the centre of a large concentration of indigenous people, was constructed and placed in operation without a good plan for the mitigation of the impact of the project on this population. The economic need to place the facility into operation immediately upon its completion caused the displaced population to be relocated without sufficient preparation of an area to receive them. As a result, the dislocation of these people has yet to be mitigated in a satisfactory manner.[18] This seemingly callous disregard for the rights of people displaced because of overriding public economic need mars the success story of these major power developments and casts a doubt on the ability of this large federal agency to continue the development of the hydroelectric potential of the basin in a socially and environmentally acceptable manner. In addition, the agency has been slow to accept the principal of multiple use of the river system and continues to operate its major facilities with little regard for the comprehensive management of the system to meet multi-purpose needs. Future optimization of the use of the São Francisco river basin will require a major change in attitude and policy with regard to this problem of integrated management.

The present and future demands for electric energy within the basin continue to outstrip the available energy. It is estimated that the demand for electric energy will double within the next 10 years. As the sites now developed represent the best sites on the river system, it can be assumed that further development of hydroelectric generation on this system will have increasingly greater impact on both competing demands for the water supplies and the riparian environment as well as severe conflicts with the existing land use. The Government of Brazil will be faced with major prioritization decisions and it can be expected that each development will face increased opposition from vested stakeholders and non-governmental organizations. Increasing need for energy for irrigation, industry, and municipal use will face trade-offs in the use of water for other demands as well as increasing pressure from those interested in environmental preservation and preservation of instream uses such as fish production and navigation. Increasing pressures can also be expected for restoration and preservation of the delta and coastal wetlands associated with the river system that are dependent upon the flood flows of the Rio São Francisco.

Navigation

From a historic point of view, the Rio São Francisco was navigable during some parts of the year in the reaches from the location of the Tres Marias reservoir in Minas Gerais to a point near Cabrobo, approximately 400 km downstream of the City of Juazeiro, Bahia. It was also navigable in its downstream reaches from a point near Pão da Acucar to the mouth of the river, a distance of approximately 200 km (map 4: Longitudinal profile of the Rio São Francisco.[19] In addition the primary tributaries of the Corrente and the Rio Grande have navigable reaches of 75 km and 350 km respectively. The primary cargo of the barges and shallow draft craft were agricultural production, livestock, and building materials. The original navigation was primarily accomplished by the private sector. However, in 1903 the State of Bahia created a public company called the Empresa de Viacão do São Francisco to manage navigation on the river within the State of Bahia. This company eventually evolved into the present Companhia de Navegacão do São Francisco, FRANAVE. However, within recent years, commercial shipping on the river has decreased to a negligible amount. From a volume of 120,000 tons within Bahia in 1987, the volume had decreased to 26,000 tons in 1994 and continues to decrease at this time. The function and existence of this public company is presently under study. With the diminished flows due to hydroelectric generation and the obstruction of the river by dams and reservoirs, major coordination and flow regulation would be required if extensive transportation on this waterway is to resume. It is estimated that, within the parameters of optimum hydroelectric generation, sufficient water could be maintained in the river to support a level of shipping that could approximate 4.5 million tons per year.[20] Such levels would require close cooperation with CHESF for optimal management of the river flow, extensive dredging and port reconstruction, and extensive investment in barges and craft. It is estimated that this investment requirement would be in the order of US$9.5bn. The waterway system or *hidrovia* would also need to connect to extensive highway and railway transportation networks to eventually get the products to the consuming market or the ports on the Atlantic. While the extent of investment appears quite high, the benefits of such a system on the economic development of this region and the value of the products

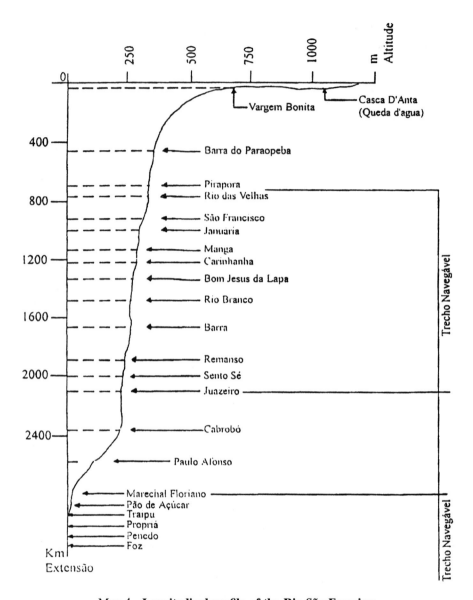

Map 4   **Longitudinal profile of the Rio São Francisco**

exported from the valley could be substantial. Consequently, the evaluation of the development of a comprehensive system of transportation for the valley that includes extensive use of the potential for water transportation should be considered.

Historic planning efforts

The valley of the São Francisco has been the subject of numerous planning and developmental effort over the years. Within relatively recent times, major planning studies were done by DNAEE[21] and CHESF with regard to the hydroelectric potential of the river.

During the 1980s a major effort was made to analyse and design a proposed project for the transbasin diversion of water from the Rio São Francisco to the non-riparian north-east States of Ceará, Rio Grande do Norte, and Paraíba as well as to the Sertão region of Pernambuco. This plan to divert large volumes of water from the São Francisco river to the fertile lands of the sertão, located in the States of Ceará, Rio Grande do Norte, Paraíba, and Pernambuco was the continuance of proposals which had originated early in the century. Initial studies, which included detailed field investigations, were started in 1981 by DNOS (National Department of Reclamation Works). In 1984, at the government's request, the World Bank financed the preparation of an Action Plan for the São Francisco Transbasin Project. This study was done by the government using an international joint-venture of consulting firms supported by cooperation from the United States Bureau of Reclamation. The main recommendations of the plan included: (i) full development of local and state water-resources management institutional capacity prior to the construction of the São Francisco diversion works; (ii) establishment of irrigation pilot areas in the plateau of Jaguaribe in Ceará and Apodi in Rio Grande do Norte; (iii) creation of a multisectoral entity to prepare detailed plans and implement the project; and (iv) the requirement that institutional constraints to efficient water use at both the state and federal level be resolved prior to project implementation.

During the past decade, the government with the support of the state governments has undertaken the construction of some of the most important hydraulic infrastructure required to exploit local water resources within the recipient states. This included the completion of the Armando Ribeira Goncalves dam in the Piranhas-Acu river in Rio Grande do Norte and the planning, design, and commencement of the Castanhão dam and reservoir to be located in the Jaguaribe river in Ceará. Recently, the federal government and the State of Ceará have started a serious effort, supported by the World Bank, to provide a solid and comprehensive legal framework for promoting the rational use of water for irrigation and other purposes.

This includes a regulatory framework that promotes the registry, allocation and use of transferable water rights. The other potential recipient states are moving to establish strong state institutional structures to provide the capacity to utilize efficiently both local water and water that might be derived from such a transbasin diversion project. The basic proposal has been debated and modified several times since its inception and is still under consideration in a greatly diminished form.

CODEVASF, with the assistance of the OAS, prepared a major master planning effort for the development of the valley which was completed in 1989. This study, termed the PLANVASF or Plano Diretor para o Desenvolvimento do Vale do São Francisco, examined the needs and potential for development in agriculture, hydroelectric development, water supply, waste treatment, commercialization of production from the valley, and the transportation needed to support production and commercialization efforts. This comprehensive planning effort forms the foundation for the ongoing efforts by CODEVASF within the valley. This plan also extensively examined the problem of the rural poor and the indigenous populations of the valley.

In 1984, as a result of the work of an Inter-State Ministerial Commission, the concept of a River Basin Committee was adopted to undertake specific planning studies within the basin. This committee, CEEIVASF, was among the first to consider the São Francisco river basin as a hydrologic unit. However, the focus of the committee was primarily technical. Although it contributed to the idea of decentralization of decision-making from the federal level to the river basin level, it lacked a mechanism for self-financing and lacked any real authority for implementation of its findings and conclusions. The work of the committee was continued through the Inter-State Parliamentary Commission for the Development of the Rio São Francisco (CIPE), which was comprised of the Presidents of the Legislative Assemblies of the five states comprising the largest portion of the land area of the basin. In addition, the local government authorities created UNIVALE, the União das Prefeituras do Vale do São Francisco, which included representation from the municipalities in the basin. This union provides technical advice on issues such as energy production, irrigation development, sanitation and human settlements, tourism, transportation, education, and environmental protection.

More recently, in 1995, the states of the north-east in cooperation with the National Secretariat of Water Resources, formed a group representing the water-resources sectors of each state to foster water-resources legal and institutional cooperation throughout the north-east including the São Francisco basin. This committee was formed in Natal, RGN with the signing of the Carta da Natal. One of the major topics of this group was the Rio São Francisco, including the consideration of the potential impacts of the proposed Transposicão Project. This group continues to meet and to discuss topics such as the water legal and institutional framework, joint efforts to study integrated management of the Rio São Francisco and other areas of potential cooperation between the states and the Federal Government in the water-resources sector.

Irrigated agricultural development

The development of the irrigated agriculture of the São Francisco basin has been a mix of the public sector and a strong private sector. The original development of irrigation for high-value crops in the basin can be attributed to the public development of the irrigation projects located in the States of Bahia and Pernambuco near the cities of Juazeiro and Petrolina. This development provided the foundation of irrigation and market potential that attracted substantial private investment. This combination resulted in the strong economy of this region. In the past 10 years, continued efforts by CODEVASF for development of federal irrigation projects in the States of Minas Gerais and Bahia have resulted in nuclei for the potential growth of agricultural economies in the outlying areas of these states. Projects such as the Jaiba and Gorituba projects in Minas and the Formoso A and Formoso H projects in Bahia form centres around which private irrigation endeavours are beginning to develop as they have in the Petrolina/Juazeiro area. While these federal projects only represent less than 10 per cent of the irrigated agriculture in Brazil, their presence represents a strong incentive for the continued expansion of the private sector. Recent proposals by the Federal Government for a "Novo Modelo de Irrigacão" or new model for irrigation, proposes strong public/private sector cooperation in the development of the agricultural sector to its maximum potential. This will probably include emancipation or privatization of the existing federal projects as well as major joint efforts between the public and private sectors to develop the major infrastructure necessary to provide incentives for private sector development.

Special Commission for the Development of the Valley of the São Francisco

The Federal Legislature, during 1995, appointed a special commission to review the planning efforts within the valley and to develop a recommended action plan to pursue the development of the vast potential of this valley. This Special Commission for the Development of the São Francisco Valley was created by Act no. 480 of 1995 of the Federal Senate, to promote discussion on strategies, policies, programmes, and priorities for the development of the valley, both present and future. Included in its mandate was the alleviation of poverty and balancing of socio-economic development and the environment in the Rio São Francisco basin, including the rehabilitation of degraded lands. This commission was empowered to undertake discussions with both private and public sector agencies and organizations in order to promote sustainable development in the basin.

The commission concluded its efforts with a final report in November of 1995 that recommended action to further the coordination of the development of the basin. During this study, the commission held hearings to accept testimony, appointed special consultants to assemble and review past studies and work with regard to the basin and reviewed the institutional framework that presently has a responsible role within the basin. This special commission was charged with:

(a) Discussion of questions over the strategy of development of the valley with particular regard for the poor of the region, and with special regard to the socio-economic well-being of the region and concern for the environment.

(b) Analysis of proposals for the management and recuperation of the environment of the basin.

(c) Constitution of a forum for the discussion over the potential economic development of the valley and the north-east region including the analysis of the potential for both public and private investment.

(d) Discussion of the potential for new projects which focus on the sustainable development of the basin.

The work of this commission represents a new and solid expression of interest in the sustainable development of the basin in a comprehensive manner. The recommendation of the commission includes the final statement that the valley of the São Francisco river represents a dynamic opportunity for modem development with advantages for the entire north-east of Brazil and that it constitutes a potential that merits the attention and priority of the entire nation. The following excerpts from the summary of recommendations of this commission provide a representative indication of the policy trends of the Federal Legislature of Brazil.[22] The commission recommendations in their entirety constitute the probable framework for the future national policy for the development of the valley of the Rio São Francisco.

Excerpts from the Report of the Senate Special Commission

The productive sector

1. The Brazilian Cerrados emerged in the past decade as an area of enviable potential for grain production. The domain of the Cerrados in the Sao Francisco Valley is the scene of accelerated growth but, unfortunately, it still requires more effective official support in the area of credit and the development of technology adapted to its particular conditions. Fledgling infrastructure is also one of the obstacles to the expansion and marketing of production. In view of this, we think it is worthwhile to make the following recommendations:
   - **Establish an Agricultural Limestone Program, with a credit line from the Bank of Brazil, in the form of investment, for a period of 2 years to stimulate the expansion of farming area and productivity gains in the Northern Cerrado – Bahia, Piauí, Maranhao and Tocantins.**
   - **Promote a Storage Program, also for the area of the Northern Cerrado.**
   - **Support agricultural diversification in the Northern Cerrado region to ensure the region's economic and productive sustainability.**
   Since the Integrated Northern Development Corridor Program already exists, created within the Bank of Brazil but with limited scope, and with the same objectives as the credit lines proposed herein, it is suggested that its name be changed to Northern Cerrado Development Program and that its geographical area be expanded (to include Western Bahia and the State of Tocantins, as well as the Cerrados of Piauí and Maranhao), modifying its scope and strengthening the availability of its financing, so that it can not only fulfill its original functions but also constitute an effective instrument for Federal Government action to develop the Northern Cerrado.
2. **Irrigated Agriculture** clearly modified the reality and perspectives of the Sao Francisco Valley. From a predominantly semiarid region, coexisting with the natural problems of periodic droughts, and confined to subsistence farming, the Valley is emerging as a nationally renowned alternative area for the production of vegetables and fruits, and is already participating successfully on world markets. However, the expansion of

233

irrigated agriculture remains affected by obstacles that limit and hinder its full development. We therefore present recommendations on each aspect of this activity.

2.1 Financing

(a) For private projects under implementation:
  - **Create a credit line in BNDES and BNB that contemplates investment plus working capital (for costing of up to two and a half crops per year), along the lines of industrial projects.**
(b) For private projects under operation:
  - **Create a specific credit line for 'irrigated costing,' over a 12-month period, through a rural credit policy carried out by the Bank of Brazil and a private network; and**
  - **Adopt the system of evolving guarantees for costing credits supported by pre-purchase contracts and other market variants, especially those related to fruits.**
(c) For public projects:
  - **Continue to make use of external financing resources, especially from the IDB and World Bank, as well as to link the financial investment corporations of these institutions (for example, IFC) to finance private agents who may participate in these projects;**
  - **Assure, both in the budget and financially, public counterpart resources for the projects, including external (such as OECF) or local (through BNDES) cofinancing;**
  - **Use the public services concessions policy (Law n° 9.074/95, art. n° 1,V) to involve the private sector in the implementation and operation of the irrigation system, as well as a means to promote the effective 'emancipation' of operating irrigated areas; and**
  - **Clearly define investments for public infrastructure, which will be made on a nonreimbursable basis, such as the implementation of primary infrastructure to transport water to the area to be irrigated.**

2.2 Costs

With the aim of reducing public irrigation costs in the Northeast, the following guidelines should be adopted, among others:
- **Transfer responsibility to states and municipalities for the implementation of social infrastructure (for example, health, education, etc.);**
- **Encourage co-participation by electrical power licensees in the implementation of transmission lines and substations, along the same lines as those operating with private projects;**

- **Assign responsibility and duties, according to government level, in relation to highway linkages; and**
- **Eliminate expenses for urban equipment, such as housing development (the irrigator should preferably reside on his own plot), airport construction, public buildings, etc.**

2.3 Management of irrigated areas

- **Make flexible the model for settlement of irrigated areas, defining on a case-by-case basis the best development scheme – ranging from exclusive settlements of colonists to exclusive business settlements. The current situation, defined a priori by decree, does not correspond to policy aimed at optimizing the productive potential of irrigated areas.**
- **Adopt strict efficiency criteria aimed at the recovery of public investment, according to previously performed feasibility studies.**
- **Establish and adopt production plans for irrigated areas or centers in order to encourage specialized production or a suitable mix of production.**
- **Encourage production and productivity gains in relation to land, labor and water in order to optimize the use of investment.**

2.4 Technology and human resources

- **Develop, in partnership with private initiative, research programs aimed at improving and disseminating technology suited to irrigated agriculture, under the Sao Francisco Valley's various natural conditions.**
- **Promote human resources training and improvement programs in the Sao Francisco River Basin, aimed at different economic activities (fruit growing, horticulture, livestock, fisheries, agro-industry, etc.).**
3. The opening of markets to the import of agricultural products, as well as the achievement of new markets and the improvement of agricultural production, imply the introduction of product quality control, which involves:
    - **Modernizing, training and streamlining disease control, either in relation to internal production conditions or to imported products.**
    - **Including in the Foreign Commerce Program a scheme for marketing and for market promotion and information, in order to support efforts already made by various producers' groups.**

Infrastructure

The infrastructure issue is at the center of the development policy for the Sao Francisco Valley, since it conditions and limits the rational use of resources, the expansion economic activities that will promote the well-being of the population and the very sustainability of development. To date, however, the Valley's infrastructure has been implemented only sporadically and disjointedly.

It is understood that it is essential not only to make up for lost time and streamline efforts but also to set into motion a process of linking the agencies and sectors involved in order to conceive and formulate medium- and long-term solutions.

1. The current development scenario in the Sao Francisco Valley needs a new stance with respect to **power** generation. In light of requirements for different uses of the basin's water resources there is a need for redefinition of priorities and greater linkage is needed between the power sector and other competing uses within the basin. We recommend the following points, among others:

   - **Consider concluded the power generation cycle along the main course of the Sao Francisco river, conditioning new uses to interest in regulating the river, navigability and irrigation;**
   - **Promote operational integration between CHESF and CEMIG, that is, among Tres Marias, Sobradinho and the Paulo Afonso falls, in order to regulate the river, minimizing the impact of floods, assuring operational flow conditions for navigability and facilitating the operation of irrigation projects. DNAEE should demand that, in the short term, both companies jointly prepare an operational plan to be reviewed by other interested parties;**
   - **Enable hydroelectric use from Formoso, on the Sao Francisco river, upstream from Pirapora, which is important for controlling floods and improving navigation conditions on the river.**
   - **Deal urgently with the conclusion of the Itaparica resettlement, in the lower-middle Sao Francisco. It is absurd that, with the hydroelectric plant operating, the living standards of the population removed from flooded areas have not yet been restored;**
   - **Promote the timely completion of the hydroelectric development projects in Sacos (116 MW) on the Formoso river, and in Sitio Grande (19 MW) on the Femeas river, both of which are tributaries of the Sao Francisco river;**
   - **Conclude the Barreiras substation, operating it at a potential of 230 KW, with works under the responsibility of CHESF, to allow oil to be replaced by electricity in the operation of nearly 400 irrigation pivots in western Bahia, representing an irrigated area of 40,000 hectares;**
   - **Implement a rural electrification program in western Bahia, through the BNDES operation-program, to allow irrigation, processing of agricultural production, implementation of agro-industries and improvement of producers' living conditions, as well as fostering regional income generation and retention.**

2. The issue of **transportation** in the Sao Francisco Valley has been one of the strongest obstacles to the marketing of production. Along with the

inefficiency of road transportation and very poor road conditions, there is a need to harness the flows of transportable production under a more modern, economical and flexible integrated scheme. In view of this, the following are recommended:

- **Implement immediately the Sao Francisco waterway, a veritable backbone crossing the region from north to south and connecting the Northeast with the Southeast of Brazil, which remains unharnessed. At present, highways, with much higher costs, remain responsible for the greater share of commodity transportation within the region. This is not efficient in rational and economic term in light of the cost-benefit ratio of the immediate operation of the waterway. The allocation of the less than US$10 million needed to transform the current navigable route into a waterway is a low-cost, highly significant investment.**
- Begin the process of enhancing the intermodal transport system of the basin, with special emphasis on the railway sector, by recovering the Central line (Juazeiro-Salvador) and implementing the Trans-Northeastern line: Petrolina-Salgueiro (PE) Missao Velha (CE);
- **Recover and restructure federal highways BR-020/242/166 and BR-135 which link productive areas with consumer markets and seaports. The waterway will also place the Sao Francisco valley on a much broader stage with clearly defined inter-regional integration roles.**
- Streamline the application of the port modernization law which will allow rather promising partnerships to be made in this sector, as long as (in the case of the modernization, expansion and construction of river ports) there are no problem which hinder the modernizaton of seaports, such as those related to the use of labor;
- **Conclude the modernization of the Petrolina (PE) airport and plan the modernization of airports in other agro-industrial centers, in line with increased local production and its consolidation.**

Water-resource and environmental management

To date, the water potential of the Sao Francisco basin has not been used in a rational manner nor have requirements been adjusted to sustainable resource maintenance and environmental preservation. These uncoordinated actions by various levels and sectors of government are not being conducted in such a way as to enable proper management and resolution of conflicts. It is therefore necessary to adopt a position regarding the river's different functions, making them compatible and assigning priorities, and keeping in mind the region's sustainable development. Along with the considerations on this issue made in this Report, and with the legislative proposal to create the Sao Francisco River Basin Water Resource Management Committee, the following are also necessary:

- **Streamline the water grant process, under the responsibility of DNAEE, preferably linking and decentralizing this activity by means of an operational agreement with state water resource agencies;**
- **As a part of the final evaluation studies for the Sao Francisco Trans-Basin Water Diversion Project, recommend that the following conditions be considered, among others:**
  - (a) Reformulate the project proposed in 1994, due to its evident unsuitability to the characteristics, water yield and demand ratios and morphologic features of the Sao Francisco river.
  - (b) By means of proper planning and execution, promote the optimized use of available water resources in the basins of the semi-arid Northeast, especially in those that would benefit from a transbasin diversion, in order to avoid the implementation of unnecessary infrastructure which become useless because of inadequate water availability.
  - (c) Carry out an Environmental Impact Study and the respective Environmental Impact Report (EIA/RIMA) which, among other aspects, emphasize the project's impact on the river.
  - (d) Promote broad discussion of the EIA-RIMA and the Project along the lines indicated in the country's environmental legislation in each affected state in various suitable forums as well as in federal agencies.
  - (e) Assess impact on areas in receiving basins (filtration, evapotranspiration, salinization, etc.) as well as economic and social effects.
  - (f) Preserve the installed capacity for power generation along the main course of the Sao Francisco river, especially at the Paulo Afonso falls.
  - (g) Preserve the river's navigability by means of a minimum year-round depth.
  - (h) Preserve irrigation conditions for 800,000 ha (calculation adopted by PLANVASF) in the Sao Francisco river basin.
  - (i) Proceed to implement a sufficient amount of actions aimed at the environmental recovery of the Sao Francisco basin, such as recovery of riparian forests, riparian ecology and vegetation.
  - (j) Define the operational, institutional and financial model, including timetables, type of management and criteria for evaluation and interruption of operations in risk and emergency situations in the Sao Francisco basin.
- **Preparation and execution of a broad Environmental Recovery and Preservation Program in the Sao Francisco Basin, involving the Federal Government (principally IBAMA, DNAEE and the Secretariat of Water Resources), and member states of the Basin. Such a broad plan should include:**
  - (a) Set-up of a management information and monitoring system; incorporation of water courses;
  - (b) Licensing and control of activities with potential impact on the Sao Francisco river's area of coverage and its tributaries;

(c) Establishment and application of uniform methodological criteria for analysis, evaluation and control;

(d) Preparation of a diagnostic of the current situation and follow-up of the evolution of the river's environmental conditions and its basin;

(e) Riparian forest and ecology recovery program;

(f) Implementation of parks, reserves and environmental protection areas;

(g) Development of a management model for linkage, integration and delegation of skills and duties in the environmental field in the basin area;

(h) The water use rights grant and control system; and

(i) Environmental zoning of the basin, with special emphasis on specific conditions in place at the source and mouth.

The development of a comprehensive plan should be the object of a request for international financing, shared by the Federal and State Governments, along the lines of the PAPP – Northeast Rural Development Program (World Bank) and may also involve sanitation works (only 5 of the 97 cities on the riverbank have sanitary sewer systems), flood control, fisheries program, reforestation, public health and formal and environmental education.

Administrative management of the basin

• **In view of inter-institutional linkage and political efforts what stands out is the need to establish a Regional Council for Coordinating Efforts in the Sao Francisco Valley, comprised of various organizations and agencies as well as State Government and private sector representatives. The Committee's creation should necessarily correspond to a Federal Government decision to give the Sao Francisco valley its deserved prominence and priority.**

The Council's agenda should include issues related to hydroelectric development and integrated operations management in the basin in order to make compatible the multiple uses of water and ensure the preservation of natural resources, the acceleration of irrigation projects, the implementation of the São Francisco waterway, the mobilization of rural credit as well as other items currently occurring in a disorganized manner. The existence of a forum at this decision-making level would also make it possible to maximize the use of available financial resources and adjust them to the area's specific characteristics.

The Council's management model and institutional characteristics are not clearly defined but should be such that the duties assigned to it can be carried out smoothly. In truth, it must contain sectoral and spatial elements so as not to lose sight of the broader objectives of sustainable development of the basin.

This Council should include specific Thematic Commissions to deal with, among other issues, credit and financing; agriculture, livestock and agro-industry; irrigation; transportation; energy, environment and natural resources; social service and action, etc.

- **Recover, both conceptually and operationally, CODEVASF's role as a development agency with a broad range of activities in the Sao Francisco valley. It is time to accord CODEVASF the status of a semi-public company in which there is room for the operational flexibility and decision-making criteria of the private sector, as a transitional stage for future privatization.**

It is foreseeable that CODEVASF's efforts are insufficient for leveraging the valley's development, even if we only consider public and official actions. This is why the first, administratively oriented suggestion was the Regional Committee to Coordinate Efforts in the Valley. But, in addition, a stronger private sector presence is essential so that the valley's development objectives can be achieved.

When considering a development agency for the Sao Francisco valley, the need for an increasing private sector initiative should be included, which should be organized in parallel with the public sector, along the lines of state or even sub-regional development agencies, as in the case of Tiete-Paraná.

Such an initiative by the private sector would greatly facilitate, in the medium term, better linkage between public and private efforts in the valley.

Finally, note should be made of the mobilization occurring in civil society and in political representation, such as the Sao Francisco SOS Movement, the Inter-State Parliamentary Commission for Sustainable Development Studies in the Sao Francisco Valley – CIPE Sao Francisco, the Sao Francisco Valley Union of Municipalities, among others, as significant indications of consciousness-raising in the Sao Francisco Valley which needs to be encouraged and strengthened.

The ideas expressed in this Report are the outcome of discussions and studies carried out by the commission and are certainly still in their infancy, needing to be further developed, aimed at clear and definitive structure. What is important at this time is to acknowledge that they are founded on principles stemming from consensus and thus, more than being recommendations and conclusions, they are the unanimous convictions of all those who study, live in, and work for the development of the Sao Francisco valley.[23]

Summary and conclusions

According to the final report of the Senate Special Commission for the Development of the São Francisco Valley (1995), the studies undertaken in the basin have never reflected a basin-wide approach and an integrated management perspective has never been applied.

In the same report, it is mentioned that there are no legal or institutional frameworks in place to administer and implement this integrated management approach.[24]

The Rio São Francisco is one of the most important resources of the north-east of Brazil. Its historic impact on the economic and cultural development of the basin is only dwarfed by its potential to enhance the future of the region if managed in an integrated and sustainable manner. One of the major accomplishments of the Special Commission of the Senate was to emphasize the participatory process by taking testimony from a large group of various stakeholders regarding the future of the valley. As mentioned above, the findings of the commission revealed the long history of sporadic and uncoordinated development within the basin as well as the distinct need for comprehensive and coordinated management through a multiparticipatory basin authority with the responsibility and authority to make critical decisions in both the development and operation of the river system. The careful development of such an institution is the key to any successful evolutionary integrated management process for the basin. This will require the cooperation and, in some instances, the lessening of the unilateral authority of several of the relatively autonomous organizations involved in the basin and a modification of their proprietary agendas to include consideration of the multi-purpose use of the basin resources in a sustainable manner. The careful development of this basin authority with the inclusion of all stakeholders in the management decision-making process is a formidable challenge that will require that all involved set aside personal agendas and concentrate on the well-being of the basin, its resources and its people. While the findings and recommendations of the commission are comprehensive and represent a policy framework for the future planning for the basin, it is worthwhile to summarize some concluding points and add a few additional considerations for the deliberation of those that will undertake this far-reaching task.

1. The key to the implementation of a comprehensive programme for the basin will be the development of a strong participatory authority for the basin that provides representation of all stakeholders and that has the decision-making authority, responsibility, and financial and political sustainability to address these problems on a long-term basis. This must be approached carefully to ensure that this does not just evolve into another non-productive layer of bureaucracy or into a politically motivated mechanism for special interests. This authority must truly represent the interests of the citizens of the

basin and be directly responsible to this public for demonstrable results.

2. The process of developing a legal, institutional, and management framework for the basin will require fairly major modifications of the legal interrelationships between the existing institutions with roles in the basin. This could be politically controversial and will require considerable political fortitude to achieve an optimum result in the face of the many special interests that will emerge.

3. The process will require a great deal of consensus building and a strong participatory approach at all levels, coupled with a well-planned public information and education programme that includes the citizens of the basin, private sector business, the media, the political and governmental leaders within the basin and the bureaucrats who influence the basin and that have vested interests in the status quo.

4. The consensus building that will be necessary can only be accomplished if full and transparent information is available to both the decision makers and the stakeholders within the basin. This will require the establishment of a comprehensive and accessible database for the basin that reflects the hydrology, meteorology, geomorphology, ecology, sociology, and economic parameters of the basin. Unless such a database is developed in a manner that gains the trust of all stakeholders, controversies over data and the impact analyses that will depend on these data will overshadow the compromises needed to reach consensus. One of the first steps must be to examine all available data and to determine where gaps in either sufficiency of data or credibility of data exist. Efforts must then be undertaken to provide the most comprehensive database possible and to reach agreement of all involved in the planning and negotiation process as to the credibility of the database for the purposes of the planning and management of the basin.

5. In order to assess adequately the impacts of operational decisions within the basin, a credible decision support model, operations model, and decision mechanism will need to be developed. This will have to be supported by historic and stochastic hydrologic and meteorologic models and databases that have the acceptance of all players. From these tools, the decision makers can then evaluate the impacts of different scenarios of operation and development on the different sectors and upon the different future alternatives for the basin.

6. The watershed management issues such as erosion control, revegetation, recuperation of the ecological system of the basin and the

minimization of pollution through point and non-point source controls, education, modification of irrigation practices, herbicide and pesticide use, industrial waste discharge control, and sewage collection and treatment must all form a part of the basic strategy for the basin. Each of these pieces is a link in the chain that will hold the sustainable management plan for the basin together.

7. In order for such a programme to be developed and to be sustainable over the long term, sustainable mechanisms for funding the management and administration of the programme must be found that will ensure that the programme does not become crippled by political indecision, changes in government or influence by special interests. This challenge alone is formidable.

8. The endeavour to develop a sustainable integrated resource management plan for the Rio São Francisco basin will not be easy and will be evolutionary. It is imperative that such planning be accompanied by realistic expectations as to the timing and practicality of results. It will take a great deal of time, money and strength of purpose. However, the rewards to the economy, culture, and ecology of the basin will be enormous. The creation of a sustainable development and management plan and programme for the natural and social resources of the basin provides an opportunity to improve the well-being of the entire north-east of Brazil as well as that of the entire nation.

Notes

1. PLANVASF. 1989. *Sintese*. Governo Brasileiro-OEA, December.
2. PLANVASF. 1989. *Plano Diretor para o Desenvolvimento*. Governo Brasileiro-OEA, Brasília, June.
3. PLANVASF, see n. 1.
4. CODEVASF. 1994. *20 Anos de Sucesso*, Ministério da Integração Regional, Companhia de Desenvolvimento do Vale do São Francisco, Brasília.
5. PLANVASF, see n. 2.
6. Relatório Final. 1995. Comissão Especial Para o Desenvolvimento do Vale do Rio São Francisco, Senado Federal, Governo Brasileiro, Brasília.
7. Ibid.
8. CODEVASF. 1994. *20 Anos de Sucesso*. Ministério da Integração Regional, Companhia de Desenvolvimento do Vale do Sao Francisco, Brasília.
9. PLANVASF. 1989. *Programa Sectorial de Energia*, Governo Brasileiro-OEA, Brasília, July.
10. Relatório Final. 1995. Senado Federal, Governo Brasileiro, Brasília.
11. Relatório Final. 1995. Comissão Especial, Senado Federal, Governo Brasileiro, Brasília.

12. PLANVASF. 1989. *Programa de Desenvolvimento das Areas Indígenas do Vale Do São Francisco*. Governo Brasileiro-OEA, Brasília, December.
13. PLANVASF. 1989. *Plano Diretor para Desenvolvimento*, Governo Brasileiro-OEA, Brasília, June.
14. Ibid.
15. Relatório Final. 1995. Commissão Especial, Senado Federal, Governo Brasílero, Brasília.
16. CODEVASF. 1995. *Projeto Amanha*. Brasília.
17. Relatório Final. 1995. Comissão Especial, Senado Federal, Governo Brasileiro, Brasília.
18. Ibid.
19. Ibid.
20. Relatório Final. 1995. Comissão Especial, Senado Federal, Governo Brasileiro, Brasília.
21. DENAEE. 1983. *Diagnóstico da Utilizacao dos Recursos Hídricos da Bácia do Rio São Francisco*, Ministério da Minas e Energia, Governo Brasileiro, Brasília, September.
22. Excerpts from the Relatório Final of the Comissão Especial Para o Desenvolvimento do Vale São Francisco, Brasília, 1995. Trans. Ms. Janice Molina, 1996.
23. Ibid.
24. Relatório Final. 1995. Comissao Especial, Senado Federal, Brasília.

References

CODEVASF. 1994. *20 Anos de Sucesso, Companhia de Desenvolvimento do Vale do São Francisco*. Ministério da Integracao Regional, Governo Brasileiro, Brasília.

Departamento Nacional da Aguas e Energia Elétrica (DENAEE). 1983. *Diagnostico da Utilizacao do Recursos Hidricos da Bacia do Rio Sao Francisco*, Relatorio Sintese. Ministério Das Minas e Energia, Brasília, September.

PLANVASF. 1989. *Plano Diretor, Sintese*. Governo Brasileiro-OEA, Brasília, December.

—— 1989. *Programa Setorial de Energia*. Governo Brasileiro-OEA, Brasília, July.

—— 1989. *Programa para o Desenvolvimento da Irrigacao*. Governo Brasileiro-OEA, Brasília, June.

—— 1989. *Programa de Desenvolvimento das Areas Indigenas da Regiáo do Vale Do Sao Francisco*. Governo Brasileiro-OEA, Brasília, December.

Relatorio Final. 1995. Comissao Especial para o Desenvolvimento do Vale do Sao Francisco, Senado Federal, Governo Brasileiro, Brasília.

9

Policies for water-resources planning and management of the São Francisco river basin

Paulo A. Romano and E.A. Cadavid Garcia

Basis for a new basin planning model

Since colonial times the São Francisco river has played an important role in the settlement of Brazil, both as the main route followed by our explorers, the *bandeirantes*, and because of its fundamental importance in communications and transportation between the various regions of the country. It has therefore become known as the "River of National Unity."

The São Francisco valley always occupied the political, economic, socio-cultural, and ecological stage not only of its own region but also of the country. For more than 50 years it has captured the attention of the government either because of its enormous hydroelectric power potential or because of the problems brought about by droughts, floods and, more recently, environmental degradation. But there was never a policy aimed at the harmonious and sustainable development of the São Francisco basin, at integrated and rational management of its natural resources, or at their conservation.

In the past 50 years, since the 1946 Constitution mandated that the Federal Government draw up and implement an overall plan to take full advantage of the economic potential of the São Francisco basin, various actions have been taken in the region. These were mainly executed by the Federal Government, in an effort to initiate and direct its development.

Nevertheless, these interventions were never properly oriented, or sizeable enough to actually lead to development, improve the living conditions of local communities and change the socio-economic profile of the region, despite favourable circumstances, such as the exceptionally good location of the basin, which is surrounded by large consumer centres, and the fact that the São Francisco valley is a major frontier for agricultural expansion.

In general, the interventions have focused on specific sectors rather than on the basin as a whole, have been sporadic in both space and time and have suffered from institutional instability. Inadequate planning, together with management unsuited to the conditions and demands of the region, has perpetuated long-standing social and economic problems and led to the emergence of new ones associated with the environment. Among the problems arising from this some-times predatory settlement by various economic sectors, mention must be made of the following:
- electric power generation, which has done little for the lands along the banks of the São Francisco, since most existing private irriga-tion schemes are oil-driven;
- the breakdown of navigation has been interrupted because of lack of maintenance of adequate conditions;
- the negative impact of irrigated agriculture on, among other things, the environment.

Administrative errors in the planning and execution of the various development projects have also had serious consequences, such as:
- badly designed and planned projects that usually took too long to execute, with resulting higher costs;
- troubles caused by political interference;
- legislative errors leading to the fragmentation of financial resources and to the authorization of too many projects at once.

Such an approach would focus on harmonizing the demands for water with the supply existing (potential, prospective, restrictive, oppor-tunities to finance, etc.), by generating human capital (an increase in the ability of society to organize and seek necessary and legitimate changes that are within the carrying capacity of the environment). This could be a new, sustainable model, technically and operationally feasible, that could advantageously replace the old one, permitting the exploitation and the needed protection of natural resources.

The new model would be based on a commitment to defend the

interests of the region and on a thorough knowledge of its potential and limitations. Production would be structured around the principles of quality, diversification, and sustainability. The strategy and actions would be based on partnerships, cooperation, decentralization, accountability, and the new paradigms of globalization: competitiveness and the advantages of association and integration. All these principles are recognized and furthered in the coordination of the National Water Resources Policy of the Secretariat for Water Resources (SRH) of the Ministry of Environment, Water Resources and Legal Amazonia (MMA).

Furthermore, the sustainable development of the region would harmonize environmental protection and preservation with economic growth. This is no easy task, since environmental preservation usually inhibits growth.

General description of the São Francisco basin

The São Francisco basin lies between 7° and 21° (latitude south) and is therefore characterized by wide differences in climate, with rainfall ranging from 350 to 1,600 mm and mean temperatures from 18° to 27°C.

The catchment area is 645,067 km², or about 7.5 per cent of the total area of Brazil. Geographically 61.8 per cent, or 389,900 km², is in the north-eastern region; 37.5 per cent, or 237,045 km², in the south-eastern region; and the remaining 0.7 per cent, or 4,188 km², in the centre-west (the State of Goiás and the Federal District). While 300,263 km² of the basin is in the State of Bahia – 47.6 per cent of the total São Francisco basin or 77 per cent of the total north-eastern basin area – only 14 per cent is in the States of Pernambuco, Alagoas, and Sergipe (figure 1).

The São Francisco valley is a distinct region within the north-east, more than half of it (56 per cent) in the extremely arid Drought Polygon. It is divided into four physiographic regions, whose characteristics are shown in table 1 and figure 1. The climatic characteristics of the basin are determined by several factors. The topography, which has a very marked effect on the distribution of mean temperatures; they are highest over the river itself and drop on both banks as the altitude rises.

The upper basin, from the headwaters of the São Francisco to Pirapora, with altitudes ranging from 600 to 1,600 m, has a humid and sub-humid climate. Summers are rainy and winters are dry; the mean

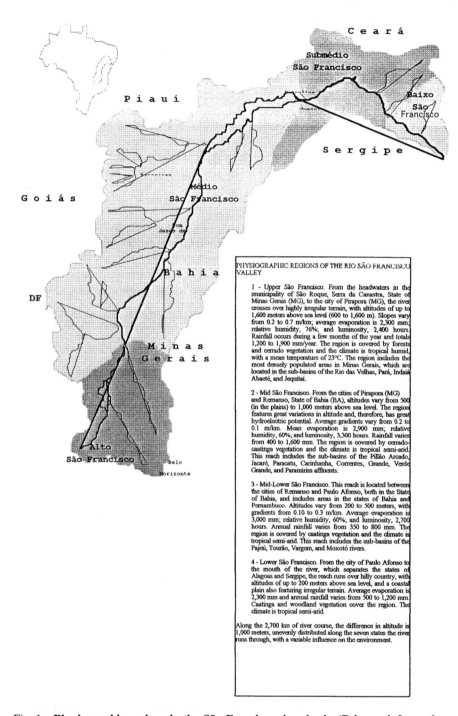

Fig. 1 **Physiographic regions in the São Francisco river basin (Primary information obtained from CODEVASF, 1994)**

Table 1 **Physical characteristics of the São Francisco river valley**

Characteristics	Upper (Canastra-Pirapora)	Mid (Pirapora-Sobradinho)	Mid-lower Sobradinho-P.Afonso)	Lower (P.Afonso-O-Atlântica)
Altitude	1,600-600	1,000-500	500-200	200-0
Wind (m/s)	SE-3	NE-4	SE-4	SE-4
Humidity (%)	76	60	60	73
Luminosity (h)	2,400	3,300	2,700	2,400
Cloudiness (0 to 10)	5	4	4	5
Evaporation (mm)	2,300	2,900	3,000	2,300
Rainfall (mm)	1,900 a 1,200	1,600 a 400	600 a 350	1,200 a 500
Rainy season	Nov. to April	Nov. to April	Nov. to April	March to Sept.
Vegetation	Forest and cerrado	Cerrado and caatínga	Caatinga	Caatinga and woodland
Climate	Tropical humid	Tropical semi-arid	Tropical semi-arid	Tropical semi-arid
Gradient (m/km)	0.70 to 0.20	0.20 to 0.10	0.10 to 0.30	3.10 to 0.10

CODEVASF (1989), supplemented by data from Portobrás, cited by Araújo.

annual rainfall ranges from 1,200 to 1,500 mm, and the average temperature is around 23°C.

The middle basin, from Pirapora to Remanso, with altitudes from 400 to 1,000 m, has a sub-humid and semi-arid climate, with summer rains. The mean annual rainfall is 600–800 mm on the eastern plateau and 1,400 mm on the westernmost edge, along the Serra Geral de Goiás. The mean annual temperature is 24°C.

The lower-middle basin, from Remanso to Paulo Afonso, with altitudes from 300 to 400 m, has an arid and semi-arid climate, with very irregular rainfall patterns and total rainfall ranging from 350 to 800 mm, depending on the altitude. The mean annual temperature is 26.5°C.

The lower basin, from Paulo Afonso to the mouth, with altitudes from 0 to 300 m, is semi-arid in the hinterland and sub-humid to humid as it approaches the mouth.

Distinct air masses moving north-east to south-west in the spring and east to west in winter and autumn. Clouds are rare, and therefore solar radiation is high.

As a result of the high mean annual temperatures, the inter-tropical geographical location, and the lack of cloud cover throughout most of the year, potential evapotranspiration is very high, varying with the temperature. The rates are highest (2,140 mm) in the sub-humid part of the basin and drop to 1,300 mm in the higher zones at the northern end.

The São Francisco has 36 tributaries, 19 of which are perennial rivers. The mean annual discharge into the Atlantic ocean is 90 billion m^3/sec of water drained from the extensive and distinct catchment regions in the basin (table 2).

The irregular distribution of the available surface water results from various factors, such as extreme variability of rainfall along time and space; severe climatic conditions in the semi-arid regions, where evaporation is very intense throughout the year; and geomorphologic elements, particularly the imperviousness of crystalline soils, which, together with the type of vegetation, intensify runoff.

The great diversity of geological formations, topographic conditions, and climatic interference results in a great variety of soils, which in turn produces three main types of vegetation, forming three district zones:

– In the upper and middle São Francisco latosols and podsols, suitable for agriculture, predominate. There are also quartzitic sands. In the mountainous areas, intermediate soils and lithosols are

250

Table 2 **Main characteristics of the catchment basin of the Rio São Francisco basin**

Rivers	Seasons Catchment Areas (km²)[a]	Mean annual discharge (m³/s) Specific yields (1/s/km²)
São Francisco	Três Marias and Pirapora	707 and 768
(downstream)	49,750 and 61,880	14.21 and 12.41
	Barra do Jequitaí and Cach. Manteira	1,015 and 1,132
	90,990 and 107,070	11.16 and 10.57
	São Romão and São Francisco	1,520 and 2,082
	153,702 and 182,537	9.89 and 11.41
	Januária and Manga	2,168 and 2,050
	191,700 and 200,789	11.31 and 10.21
	Carinhanha and Morpará	2,207 and 2,421
	251,209 and 344,800	8.79 and 7.02
	Barra and Juazeiro	2,652 and 2,731
	421,400 and 510,800	6.29 and 5.35
	Pão de Açúcar and Traipu	2,847 and 2,980
	608,900 and 622,600	4.68 and 4.79
Left bank:	Porto Alegre and Barra do Escuro	436 and 251
Paracatu and Urucuia	41.09 and 24,658	10.45 and 10.18
Carinhanha and Corrente	Junenília and Porto Novo	150 and 251
	15.32 and 31,120	9.47 and 8.07
Grande	Boqueirão	262
	61,900	4.23
Right bank:	Porto Mesquita and Várzea da Palma	140 and 292
Paraopeba and Das Velas	10,300 and 25,940	13.59 and 11.26
Jequitaí and Verde Grande	Jequitaí and Boca da Caatinga	46 and 19
	6,811 and 30,174	6.75 and 0.63

a. Depending on season.
DNAEE (1989). CODEVASF (ed.).

also frequent and associated with cerrado-like vegetation. Caatinga vegetation is found where rainfall is at its lowest.
- In the lower-middle São Francisco non-calcic brown soils, regosols, lithosols, quartzitic sands, planosols, vertisols, intermediate soils, and solodized solonetz soils predominate. This area has the basin with the least farming potential and the smallest possibility of irrigation.
- In the lower São Francisco podsols, latosols, lithosols, quartzitic sands, and hydromorphic soils predominate. The potential for irrigated agriculture depends on the topography and drainage.
Salinization of the soils can occur where rainfall is low and the water

Table 3 **Suitability of soils for irrigated agriculture in the Rio São Francisco basin (1,000 ha)**

State	Appropriate Soils (A) (A/B,%)	Soils under study	Inappropriate soils	Total (B)
Minas Gerais	10,534 (41.6)	1,175	13,608	25,317
Bahia	17,592 (54.0)	1,844	13,146	32,582
Pernambuco	1,630 (22.7)	470	5,067	7,167
Alagoas	405 (24.8)	501	725	1,631
Sergipe	150 (18.5)	127	532	809
Total	30,311 (44.9)	4,117	33,078	67,506

CODEVASF (1994).

table is close to the surface. It is estimated that 20–30 per cent of irrigated areas in the arid regions of the basin require underground drainage to maintain their productivity.

Tillable soils far exceed the availability of water. Any programme contemplating the farming of 1.8 million hectares must be part of an integrated water resources development plan, to prevent conflicts among the various sectors (table 3).

As for vegetation (CODEVASF, 1994), about 7.3 per cent of the basins is covered by dense or open forests (north-western Minas Gerais) or seasonal semi-deciduous and deciduous forests (western Bahia). Open fields covered by cerrado or caatinga vegetation account for 34 per cent and 21 per cent of the basin, respectively. Other areas consist of ecological/preservation sanctuaries (1 per cent) and reforestation projects (0.9 per cent).

According to 1996 data, the São Francisco valley contains 463 municipalities, of which 82.5 per cent are entirely within the São Francisco basin and 50.1 per cent within both the basin and the Drought Polygon. The 1991 IBGE population census shows 23.8 million people living in the 645,300 km^2 region, at a density of 37.5 inhabitants/km^2, or 53.8 per cent of the total population of the seven states comprising the basin.

In the 1960s, 4.1 million people migrated from rural to urban areas, 42 per cent of whom settled in cities. In the 1970s, another 4.7 million people migrated, 63 per cent of them to urban centres. This reveals the low and continuously decreasing retention capacity of rural areas in the north-east compared with the negative growth of its pop-

Table 4 **Evolution and trend in area irrigated, 1960–1994 (1,000 ha)**

Period	Brazil	North-east	Valley
Until 1960	461.6	28.6	10.8
Until 1970	795.8	116.0	60.2
Until 1975	1,086.8	163.4	88.0
Until 1980	1,481.2	261.4	144.5
Until 1985	1,853.7	335.8	205.0
Until 1990	2,911.7	732.5	232.6
Until 1994	–	–	300.0
Lineal trend[a]	84.0	24.9	8.5

a. Estimated.
CODEVASF (1995).

ulation: from 28 per cent in the 1960s to 16 per cent in the 1970s to negative growth rates in the 1980s. Population growth will continue to decrease in the future and will stabilize at 1 per cent in the 2010–2020 decade, when the population is estimated to be 60.6 million. Other demographic, social, and economic indicators point to the de-concentration of the urban poles of attraction, lower mortality, higher income and employment rates, and an improvement in the region's infrastructure, water supply, and basic sanitation services. The effect will be heavy pressure on water resources to be distributed among alternative user groups, competing sectors, and strategic sub-regions. Thus, conservation measures are needed to set priorities for the use of water and ensure its integrated management on the basis of the environmental capacity.

Irrigated farming has been encouraged by a series of governments, which have provided transportation, energy, and water infrastructure and built large irrigation projects (1950/60). Table 4 shows the estimated trends in the expansion of irrigation, which rose sharply beginning in the 1980s, probably as a result of the Northeast Irrigation Program (PROINE), which was executed in 1985/86. On average, the estimated growth of the irrigated area in the São Francisco basin was 8,500 hectares a year during that period.

Among the complex economic aspects of agriculture in the basin that need to be considered in planning and managing its water resources are the costs and benefits of the irrigated areas. In the São Francisco basin, it is estimated that an irrigated hectare costs from US$11,200 to US$8,900, which is high compared with other countries.

According to the 1985 IBGE agricultural census, there were 752,150 farms in the region, covering 40.6 million hectares. The growth of agriculture has taken place without an adequate or timely increase in support services. This has had an adverse impact on the effectiveness and efficiency of the region's agricultural sector.

There are also problems with research and development, mainly due to lack of credit and financing for farmers, of adequate and sufficient knowledge and technologies to bring about change, and of rural technical assistance despite the significant results achieved in certain sectors by the Brazilian Agricultural Research Enterprise (EMBRAPA) and the state research and rural extension enterprises, the universities, CHESF and CODEVASF.

Problems

The pattern of settlement, the structure and organization of production and its goods and services, and the way enterprises in the basin are run have caused a variety of problems, creating increasingly serious threats to the environment and society, and have led to conflicts among water users. An increasing number of private and governmental activities have aimed at specific sectoral and regional objectives, often conflicting and injuring to the environment and local communities.

Power generation, water impoundment for industrial and domestic use, fishing, navigation, and irrigation are examples of sectoral water uses that have not been adequately coordinated in the São Francisco basin. Land resources, such as minerals, soils and vegetation, have been exploited without due consideration of modern proper planning and management principles.

Power generation has been the main commercial use of water in the São Francisco basin. Paradoxically, however, the communities along the banks are not adequately served by electricity. In fact, hydroelectric power provides only 10 per cent of the energy used in the private irrigated areas in the middle region of São Francisco (1,917 hectares); the rest is diesel. Furthermore, irrigation has not been adequately promoted: only one in every four irrigated hectares was a government initiative. The private initiatives have not had sufficient access to credit and other necessities.

River transport has a long history on the São Francisco, but now, at a time when the region has become a major source of grain and other agricultural products, it no longer exists because of a lack of incen-

254

tives to the private sector and because of the 1,312 km of precarious navigation along the main reaches – from Pirapora (MG) to Juazeiro (BA)/Petrolina (PE), and the 2,008 km from Piranhas (AL) to the mouth.

Inadequately planned projects have caused serious silting at the mouth of the mouth. Formerly, deposits accumulated during the months of lower flow were sent to the sea during the flood months. In a 1943 report to DNPVN, Furtado Portugal stressed the need for a comprehensive rather than partial approach to the large hydro-electric and irrigation works being planned for the middle and lower São Francisco. Otherwise, he warned the government, serious problems could be expected downstream. But the works were executed without meeting the requirements set out in that report and the water quality suffered, causing negative impacts on the environment and serious problems in the outfall and estuary.

The Belo Horizonte metropolitan region is the largest and most densely populated area in the upper São Francisco basin. Some of the environmental effects of the primary, secondary, and tertiary economic activities in and around Belo Horizonte are irreversible. According to CETEC/MG laboratory results, the water of the Rio das Velhas reach, downstream from the Belo Horizonte metropolitan region, for example, has a high content of sulfate, chloride, sodium, and potassium, high concentrations of faecal coliforms and total solids, and high turbidity.

Progressive clearing of the native vegetation for farming and for making charcoal for steel mills and brick and stoneware works has also generated particularly serious negative impacts on the upper São Francisco environment. Soil loss alone is estimated at 0.17 mm/year (table 6).

In the Bahia area of the basin, the principal problems are associated with large dams, deforestation for cattle ranching, and the dumping of raw sewage into the river. In the State of Pernambuco, desertification is a serious problem, specially in the caatinga, considerably worsened by deforestation and intensive and inadequate soil management.

In the State of Alagoas most of the problems occur in the area originally covered by the Atlantic rainforest, 95 per cent of which has already been destroyed. This region is the most densely populated and has the most primary and secondary economic activities. Large distilleries, the burning of sugar-cane fields, and the use of

Table 5　Sediment production, mean annual soil degradation, and other information recorded at the sediment stations on the Rio São Francisco basin

River, station, and period	Catchment area (km^2)	Net discharge (m^3/s)	Mean annual sediment production (t/km^2/year)	Mean annual soil degradation (mm/year)
São Francisco Andorinhas Oct. 1972–Dec. 1985	13.00	248	228.0	0.14
Pará Porto Pará July 1960–Jun. 1961	11,300	145.0	44.0	0.03
Paraopeba Belo Vale Sept. 1972–Dec. 1981	2,690	43.7	582.4	0.36
Indaiá Porto Indaiá Oct. 1977–Aug. 1985	2,260	52.1	1,031.9	0.64
São Francisco Pirapora at Barreiro Dec. 1975–Nov. 1982	61,880	775.0	116.0	0.07
Das Velhas Jequitibá (Raul Soare Bridge) Dec. 1975–Nov. 1982	6,292 (4,780)	75.1 (74.4)	312.1 (661.9)	0.20 (0.41)
Das Velhas Honório Bicalho Mar. 1975–Dec. 1982	1,642	32.7	705.21	0.44
Paracatu Santa Rosa (Port Alegre) Apr. 1976–Nov. 1982 (1966/74)	12,915 (42,120)	171.0 (441.0)	154.4 (123.4)	0.10 (0.08)
São Francisco São Romão (Pedras da Maria da C.) Dec. 1968–Mar. 1975 (1972/75)	154,870 (191,063)	1,727.0 (1,981.0)	128.2 (92.6)	0.08 (0.06)
Rio Correntes Santa Maria da V (Porto Novo) May 1967–Apr. 1975 (1972/75)	28,720 (31,121)	214.0 (216.0)	18.7 (29.2)	0.01 (0.02)
Rio São Francisco Gameleira (Morpará) Apr. 1972–Feb. 1975 (1978/84)	309,540 (344,800)	2,582.0 (2,929.0)	84.5 (62.3)	0.05 (0.04)
Rio São Francisco Pilão Arcad (Juazeiro) Dec. 1968–Dec. 1973 (1967/75)	443,100 (510,800)	2,703.0 (2,666.0)	41.5 (48.6)	0.03 (0.03)
Rio São Francisco Petrolândia (Traipu) Aug. 1980–Dec. 1984 (1968/74)	590,790 (622,520)	3,454.0 (2,905.0)	32.6 (30.4)	0.02 (0.02)

Adapted from Carvalho (1994, pp. 245–247, simplified).

Table 6 **Rio São Francisco reaches and water quality**

Reaches	Class	Characteristics
From headwaters to confluence with Ribeirão das Capivaras	Special	Domestic supply with little or no treatment; preservation of the natural balance of aquatic communities
From confluence with Ribeirão das Capivaras to confluence with Rio Mombaça	1	Domestic supply after simplified treatment; protection of aquatic communities; primary contact recreational activities; irrigation of garden vegetables consumed raw and fruits growing close to the ground; natural and intensive breeding (aquaculture) of species for human consumption
From confluence with Rio Mombaça to mouth	2	Domestic supply after conventional treatment; protection of aquatic communities; primary contact recreational activities; irrigation of garden vegetables and fruit trees; natural and intensive breeding (aquaculture) of species for human consumption

Souza and Motta (1994).

pesticides pose a most serious threat. In the tropical semi-arid hinterland, the clearing of native vegetation has led to soil erosion and desertification. In the State of Sergipe, the problems of the basin are related to the sewage disposal in most municipalities and the resulting effect on public health.

According to Administrative Ordinance No. 715/89-P, dated 20 September 1989, issued by the Brazilian Institute for Environment and Renewable Natural Resources (IBAMA), which defined the federal streams in the basins, almost 10 years ago the São Francisco showed the results presented in table 5. The ordinance also specified problems demanding immediate solution before the river water could be used for domestic consumption in the middle lower and lower São Francisco.

Water should be considered an economic good. Obtaining it in adequate quantity and quality, under spatial and temporal restraints, to supply all types of uses and users, now and in the future, is subjected to increasing cost pressures. Yet there is still lacking the information needed to make water's economic, social, and ecological

Table 7 **Extreme poverty rates in the municipalities of the São Francisco basin (by state)**

State	40–50% families in extreme poverty		More than 50% families in extreme poverty	
	No. municipalities	%	No. municipalities	%
Minas Gerais	15	7.6	0	0
Bahia	46	40.3	65	57.6
Pernambuco	26	44.0	28	47.6
Sergipe	23	88.5	21	75.5
Alagoas	22	46.8	24	51.1

Peliano (1993), quoted by Araújo (1996).

importance known and to foster public awareness of its market value, with a view to enforcing appropriate integrated conservation and management.

In the Drought Polygon particularly, potable water is a scarce and vulnerable natural resource. In the large cities of the north-east, by contrast, 30–40 per cent of treated water is wasted, mainly because of a "culture of abundance." Thus, potential and even actual conflicts can be estimated on the basis of the vulnerability indicators of most temporary streams and even of some reaches of the São Francisco.

Water supply and basic sanitation are precarious and the riverside population is frequently affected by water-borne diseases. The lack of adequate and timely health care in the São Francisco basin is a frequently mentioned cause of migration to the cities (table 7).

The fishery potential of the São Francisco is still considerable, though under serious threat. Growing and uncontrolled exploitation by more than 41,000 artisanal fishermen, together with inadequate management of fisheries and large interventions (dam construction, clearing of gallery forests, industrial and domestic pollution, the effect of pesticides and mining operations, etc.), have caused changes in the composition and behaviour of the fish populations, especially of migratory species. These changes are translated into low productivity and a threat of extinction of some species as habitats are destroyed.

Per capita electric power consumption is very low among the local populations, as a result of deficient supply and low family incomes. This contrasts with the great hydroelectric potential of the Rio São Francisco, which is estimated at 10,379.2 MW/year, of which only 5,840 MW/year are in production or under construction.

Objectives

The water resources planning and management policy should be harmonized with, integrated into, and synergistically complemented by a sustainable development model. Such integration should begin with the regional plans, the 1996–1999 National Development Plan, and the National Water Resources Policy, whose implementation should include clearly defined objectives, actions, project, goals, and strategies. Some of these objectives will focus on:

- creating the necessary legal, institutional and technical conditions to harmonize multiple water uses, considering the economic, social, and ecological conditions in each region of the country and increasing water scarcity. This overall objective requires the design and implementation of a new water management model. In this line of action, the goals proposed for 1996–1999 are the following:
  - to draw up five plans for the integration of the São Francisco basins with other basins;
  - to adopt a management system for groundwater sources;
  - to create a national register of water users;
  - to train human resources;
  - to promote three educational campaigns; and
  - to increase and maintain the hydro-meteorologic network;
- increasing the drinking water supply for rural populations, on the basis of an integrated and sustainable utilization of the water potential that respects local conditions and limitations, particularly in the semi-arid region, with actions and projects such as the strengthening of the water infrastructure in the north-east (Prohidro); building community cisterns, shallow wells, underground dams and pools; and drilling, installing, and recuperating deep wells;
- fostering regional and sectoral investments that value the labour force by acknowledging its potential;
- organizing and negotiating training activities (development of human capital) and decentralization, on the basis of information about the various scenarios and conditioning factors of such development.

The topic of water resources would thus be treated within broad, integrated, harmonious, and balanced contexts that include all sectoral projects, issues, objectives, and actions, such as basic sanitation, agriculture, education, transportation, energy, and particularly social matters.

The water-resources policy for the basin sets out a number of objectives, which are considered in a document entitled "Compromisso pela Vida do São Francisco" (Commitment to the Life of the São Francisco). These objectives are also found in planning and management policy. Special mention should be made of the following:
- to define an institutionalized, integrated management model for the São Francisco basin involving the active (according to the responsibilities of each stakeholder) and timely (according to the demands and possibilities of each stakeholder) participation of the federal, state, and municipal governments;
- to identify the problems that affect the river and its tributaries by means of joint action by the various spheres of government, the public and private sectors, and civil society, in an integrated and complementary manner;
- to develop jointly a master plan for the basin and its tributaries that contemplates its integration with basins in other regions;
- to ensure continuity to studies on the diversion of the São Francisco, based on sustainability and on multiple uses and users.

The various proposals and commitments made by Brazil at international meetings, particularly at the United Nations Conference on Environment and Development (UNCED), constitute the framework for establishing the objectives of the water-resources planning and management policy.

Institutional and legal framework and development problems in the São Francisco basin

The National Water Resources Management System was established by Law 9.433, which was approved on 7 January 1997. The law stipulates the development of a National Water Resources Management Plan and the creation of River Basin Committees in order to decentralize the actions taken.

Since 1945 the São Francisco valley has received special attention from the Federal Government aimed at making use of its natural resources: first its hydroelectric potential and later soils and water for agricultural development through irrigation. For this the Federal Government created various agencies and programmes, including the following:
- the São Francisco Hydroelectric Company (CHESF), in October 1945;

- the São Francisco Valley Commission (CVSF), in December 1948, to regularize river flows, develop the hydroelectric potential, and develop agriculture, irrigation, and industry, among other activities;
- The Superintendency for Development of the Northeast (SUDENE), in December 1959;
- the Superintendency of the São Francisco Valley (SUVALE), an agency of the Ministry of Interior, in February 1967. SUVALE developed master plans, feasibility studies, and executive projects according to the standards of the World Bank and the Inter-American Development Bank. It lacked the autonomy accorded to the CVSF and after seven years was incorporated into:
- the São Francisco Valley Development Corporation (CODE-VASF). This was established in July 1974, in response to the Federal Government's need for an institution in the basin that could quickly and efficiently execute the activities of a regional development agency and serve as a link between actions of the government and of the private sector. One of the driving forces in the development of the region was irrigation, which was seen as a multisectoral undertaking that would help build a sustainable economy.

To evaluate existing basin development agencies (or create new ones), the government established several committees (some now defunct), among them the following:

- The Interministerial Commission for Studies on Flood Control for the São Francisco River, created in June 1979, under the coordination of DNOS and with representatives of CODEVASF, the regional agencies involved, and the Governments of Minas Gerais, Bahia, Pernambuco, Alagoas, and Sergipe.
- The Committee for Integrated Studies of the São Francisco River Basin (CEEIVASF), established in October 1979. In 1984 CEEIVASF set up a basin committee system, a planning system that uses river basins as planning, covers a limited number of sectoral or governmental functions, operates on a technical basis and handles almost no funds, with little legal definition, separate from the federal and state planning systems.
- The Interparliamentary Commission for the Development of the São Francisco (CIPE-São Francisco), made up of the Chairmen of the State Assemblies of Minas Gerais, Bahia, Pernambuco, Alagoas, and Sergipe, which coordinates the actions of the five assemblies in the basin.
- The Union of Prefectures of the São Francisco Valley (Univale), a joint effort of the mayors of the municipalities in the basin, with

261

vice-chairmen for energy, irrigation, sanitation and housing, tourism and leisure, navigation, education and culture, and environmental preservation.

- The Manoel Novaes Institute for the Development of the São Francisco Basin, which pursues environmental, economic, and social strategies to protect the river and promote the development of its basin. The Institute was created by the Bahia Trade Association, the Agriculture Federation of Bahia, the Federal University of Bahia, State University of Bahia, and CEEIVASF.
- The Special Commission for the Development of the São Francisco Valley, created by the Federal Senate in Requirement No. 480/ 1995 to promote wide-ranging discussion of policies, programmes, strategies, and priorities for the development of the basin. Its purposes were: to analyse proposals and projects concerned with the socio-economic and environmental balance of the basin; to analyse proposals and select appropriate means of managing and reclaiming degraded environments; to serve as a forum for discussion of the economic potential of the basin for the north-east, and consider possible public and private investments; and, to discuss and choose new projects for the region, with special emphasis on sustainable development.

Mention must be also made of the Project on the Master Plan for the Development of the São Francisco Valley (PLANVASF), which was drafted with the support of the Organization of American States (OAS) and first proposed in 1989, to guide and coordinate governmental actions and encourage private-sector activities. The plan laid out development actions to be executed and encouraged, aiming at the integrated use of natural resources. Priority should be given to increasing the production of food and raw material through irrigated agriculture and making full use of the region's electric power potential; flood prevention and control; the development of a transportation infrastructure, with emphasis on river navigation; and basic sanitation and environmental monitoring and preservation.

All government intervention efforts in the basin clearly indicate the need for integrated, long-term strategies and actions, based on sound observation and implemented according to a master plan.

Planning and management

At present the water sector is segmented in water uses and users. Actions very seldom converge and are frequently isolated, with no effective coordination following real operational principles or guide-lines that aim at the sustainable utilization of the basin's natural re-sources. The region urgently needs integrated planning and common strategies. Some of them should be directed toward eliminating obstacles to sustainable and more equitable growth in the basin; con-solidating an economy that would be more competitive, more respon-sive to international market stimuli and demands; and improving the efficiency of the economic system with attention to environmental quality.

Federal policies have traditionally favoured water supply to the detriment of a more rational water use that would enable the govern-ment to meet broader social objectives. Among the goals relating to water resources, particularly in the semi-arid areas of the north-east, special emphasis is being given to developing innovative and partic-ipative policies, such as those being defined and implemented by PRO-ÁGUA, a programme to be executed by the Secretariat for Water Resources. Priority is being assigned to the completion of irri-gation and water infrastructure works. Investments in water structure – dams, ponds, and canals – totalling R$3.5bn will increase the water storage capacity by 11.2 billion m^3. The irrigation works, covering 970,000 hectares, and the modernization of water management will require an additional R$100m. Basin committees will be established and master plans for the basins drawn up.

The water-resources policy is designed at the central level, within the new, modern government model which will be implemented at local, regional, state, and national levels, and even internationally, in the case of basins involving two or more countries.

Environment

The existing models of growth and plans for the basin are beginning to prove unsustainable and show clear signs of damaging and increasing the vulnerability of natural systems. To reverse this trend, a change is needed in the attitudes of the population and government towards the progressive depletion and erosion of natural resources, since it has become evident that the ecosystems cannot heal them-selves. A major source of pressure on water resources is the com-petition among power generation, irrigation, and domestic water supply.

Of the many complex environmental problems, only two will be mentioned here:

- the recovery and preservation of the quality status of river basins to permit the restoration of habitat, the replenishment of fisheries, and the resumption of commercial fishing, by means of sound planning and management instruments; and
- the protection of natural resources and environments, such as soils and biota, that are fast being degraded by erosion that significantly reduces the capacity of farmland and deposits sediment in rivers, where it is carried along and damages the river channel and water works equipment and diminishes the useful life of reservoirs.

Quality control and monitoring programmes are essential for ensuring water supply for human and animal consumption. Preventive measures and educational projects have proved economically sound. Water-resources planning and management action aiming at the prevention and control of water pollution from agricultural activities; the reclamation of degraded areas; the preservation and improvement of the health status of the population; and the resettlement of native populations in certain areas of the basin, among others, are essentials.

In these areas, the planning strategies aim at defining inter-agency strategies and partnerships to carry out joint activities in several time frames. In the short term:

- zoning and land use-planning studies for critical areas, in support of mitigation actions to reverse soil erosion and conserve biodiversity;
- a variety of studies to define and describe sustainable scenarios for the use and management of soils and other land resources that are under heavy human pressure;
- the setting of systematic evaluation of critical areas and other places that are subject to limitations or exposed to environmental degradation, on the basis of technical characteristics such as propensity to erosion.

In the medium and long term:

- designing and implementing reforestation programmes for the preservation areas to be reclaimed;
- identifying and implementing reclamation and protection projects for the reclamation of degraded areas;
- identifying and implementing environmental education programmes.

Sanitation and health

The following guidelines should be used in designing the basic sanitation policy for the basin:

- participation by the various local, regional, and state agents involved in the planning and management of basic sanitation services in the basin;
- relative flexibility in the provision of services, to cope with the great diversity of geographical, social, and economic situations with due respect for the local and regional characteristics, potentials and possibilities;
- the integration of actions and strategies in the sanitation sector, and their integration with those of other, related sectors and policies: water resources, irrigation, health, urban development, income/jobs, environment, etc.
- the opening of the sector to private initiative, thus combining the existing planning and management capability with the possibility of new funding; and
- the strengthening of the regulatory function of the Federal Government, and of the executing, control, and monitoring functions of the municipal governments.

Energy

Despite the high hydroelectric potential of the basin, studies have shown that the low-income population and some economic sectors, such as irrigated agriculture, are inadequately supplied. Besides, in some cases, regional sources of hydropower have been depleted.

To make full use of the hydroelectric potential of the region while dealing with the problems of water resources and environmental quality, some management measures must be planned, effective participation of the communities and appropriate consideration of the technical and scientific principles that will ensure economic, social, and environmental sustainability. Among these measures are the following:

- building the hydraulic infrastructure to capture, regularize, and distribute water and generate electric power in the São Francisco basin and other adjacent basins, in order to meet local and regional needs;
- generating power from alternative sources to supply irrigation schemes.

Transportation

The São Francisco waterway should be revitalized. Investments estimated at R$25m are designed to improve navigability between Pirapora and Juazeiro and include dredging and installing signals in some sections, such as the Sobradinho lake and the stretch between Barra and Pirapora.

Technical and economic feasibility studies should also be proposed with a view to improving navigability in some specific reaches, such as those in Rio Grande (350 km), to attract cargoes from the Barreiras region, and downstream from Petrolina, which would permit full intermodal integration of the trans-northeastern railway that ends at the Port of Suape. These improvements would significantly reduce transportation costs. Other feasibility studies are planned for the intermodal Pirapora-Unaí-Malha Centro waterway-railway connection, which offers solid prospects for development in those areas.

Agriculture and irrigation

Brazilian water policy has been centred on and directed towards physical projects, some of them questionable and many unfinished and unable to meet the construction schedule or the established goals, lacking in proper planning and no integrated, efficient resource management.

Some of the objectives of the new agricultural irrigation and drainage model are intended to support and establish technical and administrative guidelines for assessing the needs and the social, economic, and environmental (with the minimum and tolerable risk) advantages of achieving sustainable water management. SRH/MMA is preparing to perform its mission of coordinating and providing guidance to its partners in the establishment of strategies that will encourage farmers and agro-industry to use irrigation schemes that obey economic, market, and environmental sustainability criteria; establishing environmental standards and indicators of water use for irrigation purposes; and assisting, encouraging, and providing incentives for participative planning and management of irrigation schemes.

The results expected from the new model include new techniques and technologies for water tapping, use, and management in irrigated areas; the growing of quality products; agro-industry linked to primary production; marketing; other agricultural development instruments such as lines of credit, technical support to farmers, etc. These results are consistent with the planning done in sectors such as transportation, education and training, energy, research and development, science and technology, and social organization, among other proposals for the development of the São Francisco basin and its area of influence.

Education

In this sector, the diagnosis points to serious, widespread deficiencies in the educational infrastructure; in the composition, qualifications, and numbers of teaching staff; and particularly in the results achieved by the students. Losses due to school repetition and dropping out in the north-east have reached levels that are intolerable in a development context. Environmental education and vocational training are two important avenues for the development of the human capital required for the sustainable development of the basin and its area of influence.

Many activities in other development sectors, such as technology generation and dissemination, the development of agriculture and crafts, the conservation of the cultural heritage, education for basic sanitation, and the development of tourism, among others, are directly associated with education and with an awareness of the need for and advantages of "clean," sustainable processes, products, and services.

Research and development, science and technology

When these activities are properly oriented to avoid the "second generation" negative effects of modern but inadequate technologies on the environment, they permeate, influence, and drive all the sectors of development with special emphasis on education and human capital formation. The advantages of the economy and of modernity are firmly anchored in the results of these activities.

Water-resources planning and management policy should profit from the actions and strategies contemplated in the legislation establishing the National Water Resources Policy and in the Pluriannual Program of the Federal Government, which in the science and technology areas propose, among other things:

- Revising the structure of fiscal incentives that supports research and development, approving incentives for projects, and encouraging foreign risk capital in these activities;
- Strengthening the infrastructure and consolidating existing centres of excellence and human resources (social capital) for research and development and establishing technology centres to disseminate modern practices and increase technology transfers; and

267

– Providing direct support to private-sector research and to the processes of innovation in small and medium-sized companies and encouraging interactions between the private sector and the universities. Part of this effort should be aimed at adapting policies on environmental impact to criteria based on research and development and at the integrated conservation and management of natural resources in the basin and its area of influence.

Social organization

Among the priority sectors included in the planning is social for decentralization and for the setting up of partnerships and schemes of shared responsibility.

The Solidarity Community Program is in charge of the main federal actions in the social area, in coordination with the pertinent ministries. It works from a municipal approach, along two lines:
– one emphasizing food supplementation for low-income populations, which in 1994 served 150 municipalities in the basin;
– one focusing on projects in food, housing, job generation, health, and education, which in 1994 served 17 municipalities known as "pockets of property."

It must be noted that the water-resources planning and management policy includes the creation of sub-basin councils, committees and agencies, for which SRH/MMA is preparing to provide guidance and stimuli.

Recommendations

The establishment and execution of integrated programmes stipulated in the Rio São Francisco basin planning and management policies should stress the need and importance of joint, decentralized, participative actions to be implemented through proposals for interinstitutional, interdisciplinary cooperation. This process involves the negotiation of integrated planning and management.

Discussion of the paradigms for the new development model that serve as reference for the water resources planning and management activities in the basin has made clear the need for more effective participation by communities organized to assume a commitment to decentralization, to engage in partnerships and to change their behaviour and attitudes, and aware of the need for cooperation/

integration of the public and private sectors, in all areas and at all levels of government.

At the macro level and in the institutional administrative sphere, the recommendations are directed towards the following:

- the creation and maintenance of an up-to-date basin information system;
- a reorganization of the planning, budgeting, and internal control systems at the various levels of government;
- the restructuring and adaptation of the integrated administration and financial system, to draw it into the formulation of the federal and state budgets and enable it to be used as a tool for planning, management, control, and evaluation; and
- establishing technical, scientific, and operational mechanisms to analyse the technical and budgetary implications of proposals, integrated into the timetables for projects, and to monitor and evaluate the execution of the projects, to reduce the number started and left unfinished.

The Basin Master Plan should be consonant with the instruments, guidelines, and principles of the sectoral planning policies that form the Natural Master Plan and Natural Resources Plan and those of the states in the São Francisco basin. It should provide for institutional strengthening and sufficient funding to carry out decentralization and planning activities, and execution of proposals for action at the various levels of government.

References

Ansoff, H.I., Declerch, R.P., and Hayes, R.L. 1987. *Do planejamento estratégico à administração estratégica*. São Paulo: Atlas.

Araujo, J.T. Comunicação pessoal. Facsimiles No. 196/96, 12/09/96 and No. 199/96, 20/09/96, CEEIVASF-SRH.

Brasília. Congresso, Senado Federal. 1995a. *Comissão especial para o desenvolvimento da bacia do São Francisco*, vol. 1. Brasília: Senado Federal.

—— 1995b. *O papel das hidrovias no desenvolvimento sustentável da Região Amazônica brasileira. Hidrovia da Bacia do São Francisco*. Brasília: Senado Federal.

Brasília. Lei No. 9.433, 8 January 1997. Institui a Política Nacional de Recursos Hídricos, cria o Sistema Nacional de Recursos Hídricos, regulamenta o inciso XIX do art. 21 da Constituição Federal. *Diário Oficial*, n. 6, p. 470–474, 9, January 1997.

Buarque, S.C. 1994. Roteiro metodológico para a elaboração do plano de desenvolvimento da Amazonas, mimeo.

Cadavid Garcia, E.A. 1996a. *Abastecimento de água potável e saneamento básico. Aspectos econômicos na avaliação de investimentos públicos em pequenas comunidades do semi-árido do Nordeste.* SRH/IICA, Brazil. (Projeto de Cooperação Técnica BR/IICA-95/004).

—— 1996b. *Plano diretor de bacia hidrográfica: conceitos.* SRH/IICA, Brasília.

—— 1996c. *Plano diretor de bacia hidrográfica: estado da arte.* SRH/MMARHAL, Brasília.

—— 1996d. *Plano diretor de bacia hidrográfica: termos de referência.* SRH/MMARHAL, Brasília.

Campello Netto, M.S.C. 1995. *Política de recursos hídricos para o semi-árido nordestino.* Brasília: Projeto ARIDAS/Secretaria de Planejamento, Orçamento e Coordenação da Presidência da República. Uma estratégia de desenvolvimento sustentável para o Nordeste, February (GT II – Recursos hídricos).

Campos, J.N.B. 1995. *Vulnerabilidade do semi-árido às secas, sob o ponto de vista dos recursos hídricos.* Brasília: Projeto ARIDAS/Secretaria de Planejamento, Orçamento e Coordenação da Presidência da República. Uma estratégia de desenvolvimento sustentável para o Nordeste, March (GT II – Recursos hídricos).

Carvalho, N. 1994. *Hidrossedimentologia.* CPRM, Rio de Janeiro.

Castello Branco, L.C. 1996. *CODEVASF – A agência para o desenvolvimento sustentável da bacia do São Francisco.* Brasília: CODEVASF.

Cernea, M.M. 1993. "Como os sociólogos vêem o desenvolvimento sustentável." *Finanças & desenvolvimento,* v. 13, n. 4, pp. 11–14.

Companhia de Desenvolvimento da Bacia do São Francisco – CODEVASF. 1986. *Plano diretor para o desenvolvimento da Bacia do São Francisco – Síntese da etapa I.* Brasília: CODEVASF/PRONI/OEA, November.

—— 1989a. *Plano diretor para o desenvolvimento da bacia do São Francisco. Relatório Final.* Brasília: CODEVASF/SUDENE/OEA, December, pp. 113–125; 136–157.

—— 1989b. *Plano diretor – Síntese.* Brasília: CODEVASF/OEA, December, Brazil.

—— 1989c. *Programa de gestão do meio ambiente da bacia do São Francisco. Relatório Final.* Brasília: CODEVASF/SUDENE/OEA, September, pp. 26–48; 49–58 (RTP 89/66).

—— 1992. *Atividades da CODEVASF na área ambiental.* Brasília: CODEVASF, July.

—— 1994. *CODEVASF – 20 Anos de sucesso.* Brasília: CODEVASF.

—— 1996. *Programa de desenvolvimento sustentável da Bacia do São Francisco e do semi-árido nordestino. Síntese.* Brasília: CODEVASF, June.

Conferência das Nações Unidas Sobre o Meio Ambiente e Desenvolvimento (CNUMAD/ECO-92) 1995. *Resolução No. 44/228 da Assembléia Geral da ONU (22/12/89). Agenda 21.* Brasília: Câmara dos Deputados, Coordenação de Publicações (Séria ação parlamentar, 56).

DNOCS. 1959. *O problema nacional das secas.* Departamento Nacional de Obras contra as Secas, *Conselho Nacional de Economia,* v. 19, n. 3 (Bulletin 3).

Luz, L.D. "O estabelecimento de cenários no planejamento do uso dos recursos hídricos." In: *Desenvolvimento sustentável dos recursos hídricos. Gerenciamento e preservação,* 3, ref.: ABRH/APRH. XI Simpósio Brasileiro de Recursos Hídricos e II Simpósio de Hidráulica dos Países de Língua Portuguesa, pp. 33–38.

Ingelstam, L. 1987. *La planificacion del desarrollo a largo plazo: notas sobre su esencia y metodologia,* n. 13. Revista da Cepal.

Nobre, P. 1994. *Clima e mudança climática no Nordeste.* Brasília: Projeto ARIDAS/ Secretaria de Planejamento, Orçamento e Coordenação da Presidência da República. *Uma estratégia de desenvolvimento sustentável para o Nordeste,* September (GT I – Recursos naturais e meio ambiente).

Nou Edla A.V., and Costa, N.L. da. 1994. *Diagnóstico da qualidade ambiental da bacia do rio São Francisco. Sub-bacias do Oeste Baiano e Sobradinho.* Rio de Janeiro: IBGE.

Peliano, A.M. 1993. *O mapa da fome III; Indicadores sobre a indigência no Brasil,* n. 17. Brasília: IPEA. (Calcula para o Vale do São Francisco/CEEIVASF.)

Projeto Aridas. 1995. *Nordeste: uma estratégia de desenvolvimento sustentável.* Projeto Aridas, Brasília.

Queiroz, J.F. 1980. "Tecnologia da aridez – a difícil convivência com a natureza adversa." *Interior,* v. 4, n. 31.

Rocha, G. 1989. *O Rio São Francisco. Fator precípuo da existência do Brasil,* 3rd. edn, pp. 13–24. São Paulo: Nacional.

Rodrigues, M.M. 1996. *Retomando o planejamento: o Plano Plurianual 1996-1999,* n. 5, pp. 3–30. Rio de Janeiro: Banco Nacional de Desenvolvimento Econômico e Social (BNDES).

Silva, W. Dias. 1985. *O Velho Chico – sua vida, suas lendas e sua história,* pp. 49–53; 227–239. Brasília: CODEVASF.

Silva, J.G. 1988. *A irrigação e a problemática fundiária do Nordeste,* pp. 18–25. Instituto de Economia – UNICAMP/Proni, Campinas.

Souza, R.O. de and Motta, F.S. 1994. *Qualidade e conservação da água, com vistas ao desenvolvimento sustentável do semi-árido nordestino.* Projeto Aridas/Secretaria de Planejamento, Orçamento e Coordenação da Presidência da República. *Uma estratégia de desenvolvimento sustentável para o Nordeste,* September, Brasília (GT II – Recursos hídricos).

Souza, A. 1996. *Programa de fortalecimento do setor pesqueiro e desenvolvimento da aquicultura.* Brasília: CODEVASF (Preliminar).

Sudene. 1995. *Sudene: uma parceria de sucesso da bacia do São Francisco. Cem anos de Petrolina.* Recife: Sudene.

10

Addressing global environment issues through a comprehensive approach to water-resources management: Perspectives from the São Francisco and Plata basins

Alfred M. Duda

Introduction

Twenty-five years after the Stockholm Conference on the Human Environment the world still faces a wide variety of critical environmental threats that have global implications: degradation of soils, water, and marine resources essential to increased food production; widespread health-threatening air and water pollution; global warming that could disrupt weather patterns, cause floods or droughts, and raise sea levels everywhere; loss of habitats, species, and genetic resources, damaging ecosystems and the services they provide. Five years after world leaders renewed their pledge at the Earth Summit in Rio to comprehensively address these issues, environmental conditions remain about the same in the North as well as the South despite greater understanding of the gravity of the situation. Wasteful patterns of overconsumption continue in many countries of the North and rapid population growth and natural resource depletion continue in many countries of the South.

It is a time of unprecedented global change. Human activities are now a major factor in overriding natural changes in species, land cover, and ecosystems that are brought about by climate variations over thousands of years. Fuelled by the triple pressures of over-consumption, globalization of trade, and population growth, various uses of the earth's surface (agriculture, forestry) have altered the land cover of large regions and endangered plant and animal habitat. Fire, grazing, and tillage bare the earth, erode the soil, deplete the land, and degrade freshwater and marine systems to the point that regional hydrologic and geochemical cycles are disrupted. Non-indigenous species introduced by humans crowd out natural species; industrial pollutants fill the skies and waters with poisonous sub-stances, and greenhouse gases as well as ozone-depleting substances change our atmosphere and climate. We are overfishing the oceans and converting coastal wetlands to diked agricultural uses which fur-ther reduce ocean fish stocks.

In the 1970s, these problems seemed local in extent. By the 1980s, research demonstrated expansion of the problems to cover entire multi-country regions. By our decade of the 1990s, there is no ques-tion that these issues are global in significance and ultimately have adverse implications for the health of the global environment. For the first time in our planet's history, mankind is accelerating global changes to the point that economic systems may collapse, human survival becomes much more expensive, and sustainability of the world's environment – our life support system – is placed in jeopardy. What can be done about it?

This paper outlines the importance of global environment threats and global pressures that are becoming significant to mankind. It is not only rapid population growth and overconsumption that create these pressures but also globalization of trade, world markets, agricultural subsidies, multinational private sector investment, and international finance institutions contribute to these environmental threats. Climate change (resulting from emissions of greenhouse gases), loss of biological diversity, and degradation of transboundary water bodies represent symptoms of increased pressure on the global environment.

World leaders began coming to grips with these threats when they took the first modest steps forward in 1992 at the Earth Summit in Rio. Various global conventions, agreements, and blueprints for action resulted from the summit and some are described in this paper to illustrate cross-sectoral linkages and implications for water re-sources and their management. Among these accomplishments was the creation of the Global Environment Facility (GEF). The new and additional finance through GEF represents an opportunity for nations to incorporate actions addressing these global issues into their development plans and are outlined in the paper.

The São Francisco and Plata basins are quite representative of challenges facing the world in addressing these global environment issues. Very complex environment–water linkages are discussed for both basins in this article to illustrate how difficult the situation really is and how unprecedented the challenge that the world community faces. Both the North and the South have enormous changes to make to meet these challenges. With cities such as Bele Horizonte and Buenos Aires, with sophisticated irrigation projects such as those near Petrolina, and with significant industrial pollution, the two river basins embody not only issues common to the South but those typical of the North as well.

Given the enormous environmental, social, economic, and political implications of these challenges, the dwindling amounts of official development assistance, and the political reality that spending in the South will focus on activities with domestic benefits rather than global benefits, how can countries respond to their new duties and obligations? The challenge is even more fundamental than that. Each global convention or action programme has its own sectoral implications, its own separate political champions, and its own fragmented approach. Which deserves priority? How can nations respond when the rules are changing as institutions evolve? How can common sense be brought to complex, confusing, changing global approaches that are made more fluid by pressures related to privatization, private sector speculation, and globalization of economic systems? There are no models for making such complex changes during times of uncertainty. What is certain is that global change is occurring on many levels and the North as well as the South has only a short time to respond before irreversible changes may diminish the quality of human existence and the sustainability of our life support ecosystems.

It is the thesis of this paper that a practical approach to making these changes can be implemented through adaptive management strategies based on a more comprehensive approach to water-resources management river basin by river basin in each country. While some interventions (such as for energy use efficiency) will need to be implemented sector-wide, some climate change, many biodiversity, and most land degradation and transboundary water resources interventions will be site-specific, will need to be selected based on countries setting priorities with participation by civil society, and will vary from one basin to another. With this site-specific nature and with the opportunity to multiply benefits depending on the location or the policy change in question, a more comprehensive, cross-sectoral, or holistic approach to basin management that incorporates opportunities for mainstreaming global environmental interventions may be attractive to donors and recipients.

Both the World Bank in its Water Resources Management Policy and the GEF recognize the importance of such a comprehensive approach. This paper makes the case for using such an approach river basin by river basin as a pragmatic platform upon which interventions may be anchored during these times of global change. Examples of opportunities for incorporating global environment considerations in improved, more comprehensive management of the São Francisco and Plata basins are included in the paper. When combined with programmes of experimenting (learning by doing) in each country as part of flexible, adaptive institutional arrangements, this comprehensive approach may provide a vehicle during the near term for nations to adapt little by little to global changes by making sensible priority interventions that might have multiple benefits in the context of planned economic development. The GEF is playing a catalytic role as recipient countries are testing different interventions and learning from the experiences. Over time, enough pilot projects, demonstrations, and policy changes will have been made so that results can be mainstreamed into development-financing strategies, and, ultimately, the goal of socially and environmentally sustainable development can be achieved.

Global environment threats and pressures

The earth's forest land is shrinking, the deserts are expanding, and soils are eroding. At the same time that the hungry planet needs more food to feed its rapidly expanding population, more land is going out of production from soil erosion (6–7 million ha/yr) and waterlogging/salinization (1–2 million ha/yr) than is being placed into production. And downstream, the eroded topsoil fills in reservoirs, canals, and rivers, exacerbates flood damage, and leads to reduced hydropower production or irrigation potential. In the US alone,

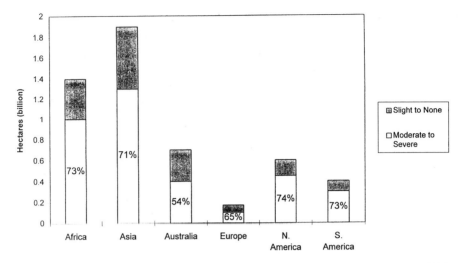

Fig. 1 **Severity of degradation in drylands (by continent) (UNEP revised 1992. Based on survey of best national estimates)**

over $5bn in damage is estimated to be caused each year from soil erosion (Clark et al., 1985).

Land degradation

In the drylands, in particular, terrible consequences are suffered by the poverty-stricken population as a result of land conversion and degradation. Deforestation, degraded rangelands, depleted soils, salinized land, and depleted aquifers impair the lives of 100 million of our planet's poor and threaten another 800 million people. Figure 1 shows that dryland degradation is a moderate to serious concern on every content, including Latin America.

Deforestation has become a globally significant phenomenon. While conversion to other uses is intended to support human existence, it may have the opposite effect of reducing the capacity of man to inhabit certain areas. While deforestation has increased dramatically over the last 20 years in the moist tropics, it is now decreasing in some Asian countries because they have run out of forests (Houghton, 1994). And the converted land is being abandoned at a record pace as its short-lived productivity declines. At current rates of deforestation, tropical forests may disappear from the earth in our children's lifetime (Houghton, 1994).

Biodiversity loss

With the loss of forests, the accelerated siltation and pollution of waterways, and the conversion of wetlands to agricultural uses, the earth's biological diversity is disappearing at unprecedented rates. Tropical forests are the focus of concern about the extinction of species because they contain so many species (perhaps 50–90 per cent of all life on earth). While the US and Europe have lost over 90 per cent of their original forest cover, Central America, South-East Asia, and West African tropical forests share the same fate. Perhaps thousands of native species may have been lost and tens of thousands of others threatened. The Global Biodiversity Assessment (UNEP, 1996) describes the nature and significance of the loss. Enormous potential economic value for pharmaceutical products and other items is at risk along with the potential improvement in quality of life that could accrue to the poor who might have become beneficiaries. Neither the ethical nor ecosystemic value of biodiversity is well understood, but once a species is lost, it is lost for all time. When many species are being lost, real concerns should be raised about the future of evolution and what this means to mankind.

Climate change

A linkage also exists among deforestation, land degradation, and climate change. Fully 25 per cent of the increase in carbon dioxide in the atmosphere can be attributed to deforestation (Houghton, 1994) and additional amounts to agricultural activities. Between 1950 and 1980, fossil fuel use worldwide increased by a factor of four. Energy use and patterns of consumption in the North fuel the concern about climate change. However, within a decade, emissions of greenhouse gases from developing countries will overtake the total emitted by the rich countries as energy use in Asia skyrockets in response to population increases and greater affluence.

Climate change threatens to disrupt weather patterns and raise sea levels around the world and to cause untold economic damage. Drastic changes in agricultural production, rainfall, and storm patterns are forecast. Entire island nations may disappear under modest half metre rises in ocean levels, and changes in storm patterns related to El Niño and the low pressure phenomenon off the coast of Argentina (associated with coastal flood damage) will accelerate damage. While the North has less poverty and more affluence available to adapt to climate change, the South with its greater poverty and lower economic capacity is more highly vulnerable to climate change and disruptions in civil society.

Cross-sectoral linkages to water problems

Water specialists recognize that these three global phenomena along with rapid population increase have an ultimate effect on local, regional, and maybe even global hydrologic and geochemical cycles. The effects will be felt in drought, floods, economic damage, increased costs, and damage to the environment. Enormously complex linkages exist among rivers, their watersheds and oceans with regard to the ability to serve as carbon sinks, preservation of key biodiversity that makes possible sequestration of carbon, and energy/weather patterns. The linkages are poorly understood and occur on so many levels that awaiting precise documentation may not be prudent. Preventive action is needed because the driving forces for adverse changes will have been set in motion and ecological as well as economic disasters may result that are international in character.

A good example involves the natural warming of the Pacific ocean off Peru, which has periodically created global changes in climate, rainfall, food production, and fishing. This effect has been called the El Niño Southern Oscillation (ENSO or El Niño for short). While this periodic ocean warming that builds storms and disrupts wind patterns has occurred on average every four years for less than a year since first noted in 1870, the effect is getting worse and is lasting longer. In fact surface ocean temperatures have increased the last several decades by 0.5°C which is the trigger for ENSO. Has climate change and ocean warming already occurred?

As noted by Tibbetts (1996), the worst ENSO occurred during 1982–1983 and the most recent serious one lasted for four years (1991–1995). During these events, droughts plagued North-East Brazil, India, Sri Lanka, China, and Africa. Crops failed in Indonesia and the Philippines. In Queensland, Australia, the drought took a $1bn toll on crops. Influences in Western Canada and the US are also reported along with large increases in rainfall in the Plata basin (Anderson et al., 1993). Fisheries collapse in certain areas of the world and explosions of algae blooms known as red tides result in depletion of oxygen and generate toxic effects (Tibbetts, 1996). As noted later in this paper, the possibility arises that devastating billion dollar floods in the Plata basin and devastating droughts in North-East Brazil (including the São Francisco basin) could be linked to climate change disruptions in ENSO and that the increase in frequency, severity, and duration of ENSO has either already resulted from climate change/or may get worse in the future. Taking this one step further, diversions of water from the São Francisco to combat droughts in Brazil's north-east may also be driven by these subtle changes in climate, and environmental damage to the São Francisco delta from additional diversions to drought areas may also then be driven by climate change.

Transboundary water resources

The world's water resources are under enormous stress, and the ecosystems, people, and economic development that depend on these resources face an uncertain future. The oceans have been fished out; estuaries have become eutrophic; and coastal zone wetlands have been drained, paved, farmed, or converted to aquaculture. Marine mammals and aquatic birds have become laden with toxic chemicals; diversions of water for agriculture have dried up rivers and lakes; pollution discharges have created health problems; and groundwaters have been overpumped and contaminated. Progress in addressing these issues has been disappointing within single nations, and resolving such concerns among nations in transboundary situations often seems impossible. Little by little, regional seas, shared river basins, international lake basins have become degraded and benefits they have historically provided are not ensured into the future. Little by little, the regional problems grow into larger problems to the point that they become globally significant.

There is no question that water scarcity and unsustainable water use threaten development in many countries. With projected population increases, many countries face dire predictions for near term water scarcity as described by Postel (1992). Competition for water for farming, human consumption, and industry is already keen and environmental needs are often ignored. Water scarcity means that food security is threatened and combined with reduced productivity from waterlogged and saline land as well as reduced productivity in eroded uplands; a strong case has been made for upcoming worldwide food shortage (Brown, 1996). Already, tens of millions of people have been turned into "environmental refugees," tens of millions more have migrated to large cities from the countryside, and the number of refugees may approach 100 million in the next century (Fell, 1996).

Worldwide, the linkage of freshwater basin activities to degradation of marine ecosystems is becoming apparent as pollutants from inland areas degrade many regional seas. Problem water bodies such as the Baltic sea need $20–30bn in remedial measures to address pollution and habitat (Kindler and Lintner, 1993). Even more will be needed to restore the eutrophic dead zone in the Gulf of Mexico caused by Mississippi river pollution or to reduce contamination in the Volga that is plaguing the Caspian sea (Wolfson, 1990). It is clear that these transboundary water-resources issues also involve single country basins that drain to multi-country marine waters because of interconnections between land activities and downstream ecosystems.

These interconnections are very complex with multiple causes of ecosystem degradation and subsequent loss of jobs and economic impacts. It is not just pollution or conversion of important wetland habitat, such as loss of mangroves due to conversion to shrimp ponds in Asia, that creates transboundary conflicts in marine waters. Overfishing is an enormous problem due to open access without adequate management regimes, overcapacity in fishing fleets with modern technology and government subsidies creating distortions in markets. All 15 oceanic fisheries are being fished at or beyond capacity and 13 are in a state of decline, some possibly lost forever (Brown, 1996). Almost one billion people rely on fish as their primary source of protein. With the oceans being fished out and irrigation expansion being stalled by costs and adverse environmental impacts, there is a serious threat to food security in the near future that has transboundary water implications.

Table 1 outlines the broad array of countries that share international rivers, lakes, and enclosed seas. Transboundary water-resource problems, conflicts, and disputes in these basins are projected to grow, as noted by Duda (1994), and the magnitude of the water crises that will develop may constitute an unprecedented, globally significant problem (Biswas, 1994).

Global pressures

Overconsumption in the North is a key factor along with deforestation in creating risks from climate change. Soon energy consumption in the South will dwarf the North's contribution. While

Table 1 **Major transboundary rivers, lakes, and enclosed sea**

International rivers	Number of countries	Lake or sea	Number of countries
Danube	14	Mediterranean sea	18
Niger	10	Black sea	17
Zaire	9	Baltic sea	9
Nile	9	South China sea	9
Rhine	8	North sea	9
Zambesi	8	North sea	9
Amazon	8	Caspian sea	6
Mekong	6	Lake Chad	6
Elbe	5	Lake Superior	2
Plata	5		
Ganges	4	Lake Victoria	5
Colorado	2	Lake Tanganika	4

Duda (1994).

some may blame these global changes simply on population growth and the poor cutting forests, the case is not that simple. Very large and unprecedented changes in the earth's vegetative cover suggest root causes that are global rather than local.

The information revolution, globalization of trade, agricultural export markets, the desire for countries to increase exports to gain foreign exchange to repay international loans, and common government subsidies result in policies that create environmental pressures. For example, recent estimates by the Worldwatch Institute document show over \$500bn a year of taxpayers' money worldwide is spent by governments to subsidize deforestation, overfishing, and other environmentally destructive activities that add up to fuelling global change (Roodman, 1996). Other examples are provided in the discussions on the Plata and São Francisco basins.

Global institutional arrangements

Protecting the global environment requires a great deal of international cooperation. Protecting the global commons requires global action, not fragmented action in a few nations. The world community made a good start at Rio as a consensus began to develop on international agreements. A variety of new conventions, agreements, action programmes, policies (such as the World Bank Water Resources Management Policy) and funding sources such as the new

281

Global Environment Facility (GEF) provides opportunities for nations to take damaged basins with significant conflicts and environmental degradation and work through to sectoral interventions in a comprehensive manner that can be more consistent with environmentally sustainable development. A brief outline of each relevant instrument follows.

Climate

The United Nations Framework Convention on Climate Change (FCCC), which became effective in March 1994, was an international acknowledgment that change in the earth's climate and its adverse effects are a common concern of humankind, and it calls for the widest possible cooperation by all countries. While recognizing that various actions to address climate change can be justified economically in their own right and help in solving other environmental problems, the convention agreed on the need for all countries, especially developing countries, to have access to resources to achieve sustainable social and economic development. As developing countries progress toward sustainable development and their energy consumption grows, they will have to consider ways to achieve greater energy efficiency and control greenhouse gas emissions, including how to apply new technologies in ways that are economically and socially beneficial. Box 1 provides a capsule summary of the FCCC.

Biodiversity

Beyond the ethical and aesthetic aspects of protecting biodiversity, loss of species and habitat have serious social and economic costs. Biodiversity loss represents lost options for being able to adapt to change, and in fact it represents an important part of the earth's natural capital.

Biodiversity is being destroyed by humans at an alarming rate and without urgent action, future options will be reduced. While maybe 1,000 species have become extinct during the last several centuries, loss and conversion of habitat worldwide have probably committed tens of thousands of more species to extinction (UNEP, 1996). Box 2 summarizes the Convention on Biological Diversity (CBD) signed at the Earth Summit in 1992.

Box 1 **UN Framework Convention on Climate Change (FCCC)**

While climate has changed naturally in the past over long periods, the accelerated pace at which greenhouse gas-included climate change may take place in the future could be too fast for some ecosystems to adjust and too expensive for some countries to adapt. This rapid pace of change was instrumental in over 150 countries signing the FCCC at the Rio Earth Summit in 1992. More than 160 countries have now become Parties to the Convention, the objective of which is to stabilize greenhouse gas concentrations in the atmosphere at a level that would prevent dangerous anthropogenic interference with the climate system (Article 2).

Various obligations of countries are listed in Article 4, including: (a) financial commitments for developed countries, (b) development of emission inventories, (c) formulation and implementation of national programmes containing measures to mitigate climate change, (d) promotion and cooperation in the development, application, and transfer of technologies, practices, and processes that control, reduce, and prevent man-made emissions of greenhouse gases.

In addition to national programmes, technology development and reporting of emission inventories as well as national steps taken to implement the Convention, Parties are required to take into account, where feasible, climate change considerations in relevant social, economic, and environmental policies and actions. For these changes to be made, the concept of sustainable development must evolve and be put into practice. Further negotiations are needed on a protocol to the FCCC that sets emissions reduction goals because it is simply a framework convention. Until the Conference of the Parties decides on permanent financial arrangements, the Global Environment Facility (GEF) operates the interim funding mechanism for developing countries to conduct projects and activities under the Convention.

Land degradation

Prevention and control of land degradation, primarily desertification and deforestation, are critical to achieving sustainable development. However, the environmental and economic consequences of land degradation are not confined to the countries where it occurs. Its impacts, in terms of loss of biodiversity, reduced atmospheric and

---

Box 2 **Convention on Biological Diversity (CBD)**

Like the FCCC, the CBD was negotiated in various sessions leading up to the Earth Summit in June 1992, where over 100 nations signed the treaty. As of late 1996, almost 170 countries had signed the Convention. The CBD reflects growing consensus that biodiversity conservation and management is an integral part of sustainable development. Many sectors of a country's economy depend directly or indirectly on the diversity of various ecosystems, and poor people, in particular, utilize these resources to provide livelihoods for their families.

The Convention's objective is the conservation of biological diversity, the sustainable use of its components, and the fair and equitable sharing of benefits stemming from utilization of biodiversity, including genetic resource. The CBD focuses on development issues which include integrating biodiversity concerns into national decision making, utilizing socio-economic policies and incentives for making decisions through environmental and social impact assessments, and controlling processes and activities threatening biological diversity. Each party has a responsibility for conservation and sustainable use of its own biological diversity through its own processes, activities, and institutions. Implementation of the Convention is through national strategies, plans, and programmes that integrate with decision-making in sectoral activities. Countries are asked to establish priorities and then report them to the CBD Secretariat through strategies and plans.

The Convention includes articles dealing with genetic resources, capacity and institution building, access to technology transfer and biotechnology. The rights and contribution of indigenous communities to sustainable use is explicitly recognized and equitable benefit sharing is to be included. Developed nations are to provide financial resources to enable developing countries to implement the CBD and benefit from it. As with the FCCC, the GEF has been named to operate the interim financial mechanism for the CBD until a permanent mechanism is chosen.

---

subterranean carbon sequestration, and pollution of international waters can be significant and global. The environmental problems of drylands cause more misery than those in any other part of the globe. In the last two decades there have been at least 10 million environmental refugees within and from the drylands. A fifth of the world's population is affected to a lesser, though often critical degree. Most of the world's poorest countries are in the dry parts of the world.

Dryland environmental problems develop for a multiplicity of reasons: drought; desiccation; land degradation; civil unrest; international conflict; economic pressure; demographic pressure; and many others. It is seldom, if ever, easy to discover the relative roles of the different factors. It is even difficult to choose measures of distress, and most, when chosen, are found to have very little geographical or temporal generality. The complexity, as well as the extent of dryland environmental problems has hampered treatment. Despite the situation the world community came to a consensus on action, as is noted in Box 3.

Transboundary waters

With respect to international freshwater basins (including rivers, lakes, reservoirs, and transboundary groundwaters), no single binding legal instrument articulates a global consensus on sound use, conservation, and development of the resources. However, a large number of bilateral and multilateral agreements and management authorities exist. In addition, the non-binding Dublin Statement and draft articles undertaken by the International Law Commission (ILC) on the Law of Non-Navigational Uses of International Watercourses (Biswas, 1994) represent some measure of international consensus.

The architecture of marine agreements, in particular, is quite complex. The marine agreements are consistent with and operate within the legal framework of the 1982 UN Convention on the Law of the Sea (UNCLOS), which entered into force in November 1994. It provides a global framework for the protection and management of the marine environment and its living and non-living resources, as is noted in Box 4. It is reinforced by a web of global and regional agreements, including those on regional seas, pollution from land-based sources, wetlands, protected areas and species, fisheries, hazardous substances, biodiversity, and climate. Agenda 21 recognized UNCLOS as "the international basis upon which to pursue the protection and sustainable development of the marine and coastal environment and its resources."

While UNCLOS refers to the linkage of freshwater basins and their influence on marine waters, little in the way of substance was negotiated to produce action. These provisions are quite controversial and

Box 3 UN Convention to Combat Desertification

The environmental problem of land degradation in arid, semi-arid and dry sub-humid regions attracted world attention during the 1970s when Africa experienced serious droughts. An approach toward international action was taken in 1977 at the UN Conference on Desertification in Kenya with the formulation of a Plan of Action to Combat Desertification. As with other international "action plans" without financial mechanisms, funding was inadequate, progress was disappointing, and a new initiative was in order.

The issue of desertification was placed back on the international environmental agenda at the 1992 Earth Summit. A key chapter of Agenda 21 addressed the global problem and recommended further political attention. This was achieved in 1994 in Paris when 102 nations signed the Convention, which entered into force in December 1996. The objectives of the Convention are to combat desertification and mitigate the effects of drought in countries experiencing serious drought or desertification. It is designed to be implemented through National Action Programs (NAPs) prepared by national consultative committees charged with the support of international cooperation, consistent with Agenda 21.

As part of the Rio process, governments recognized that desertification is not just a technical problem but rather is a social and political problem as well. The necessity for community involvement and commitment to actions on the ground are included in the text and recognize the importance of grassroots groups and NGOs in helping people who depend on drylands for their livelihood. The Convention is also to be implemented through regional action programmes contained in four annexes. Latin America and the Caribbean constitute one of the four regions.

All parties are to adopt an integrated approach addressing the physical, biological, and socio-economic aspects of desertification and drought, including paying attention to the influences of international trade, marketing arrangements, and debt. With its bottom-up approach, through community involvement, parties are asked to improve living conditions for the poor trying to maintain a sustainable existence in drylands and are called upon to facilitate participation and address root causes of desertification through changes in legislation and policies. Developed country parties are obliged to provide financing and other forms of support through bilateral and multilateral financial mechanisms to mobilize and channel "substantial financial resources to affected developing country parties."

Box 3 **(cont.)**

Unlike the UNFCCC and the CBD, the GEF is not asked to operate an interim financial mechanism and no new fund is established for combating desertification. Rather, improved management mobilization and coordination of existing funds is promoted through establishment of a "Global Mechanism." An organization is to be chosen to host this mechanism and will be directed by the Conference of Parties to promote the availability of financial mechanisms and to consider ways of facilitating funding at the national, regional, and global levels, including resources needed for transfer of technologies on a large scale to developing country parties.

---

the Global Programme of Action described in Box 5 for addressing this concern contains no funding mechanism.

The goal of negotiating conventions is to achieve agreement among countries to constrain their actions, often with perceived adverse economic implications. The political reality is that sovereignty issues – especially between developed and developing countries – will remain an overriding concern, and many countries will want to retain formal control over their policies. Domestic policy and actions are often seen to be separate from international policies and actions. However, for water quality, quantity, and ecosystem concerns, changes are often needed in each country's domestic policies and activities, including changes in sub-national (provinces, states) sectoral policies and activities. With site-specific actions being needed to address the particular problems of each basin or marine ecosystem, global conventions or regional framework conventions may not be sufficient to address transboundary water-resources issues. And with fragmented approaches for each convention, which one is priority? Where can a country begin? How can funding priority be justified?

Global Environment Facility

Two years before the Earth Summit, the Global Environment Facility (GEF) was established as a pilot programme to test new approaches and innovative ways to respond to global environmental challenges in its four focal areas of climate change, biodiversity conservation, ozone depletion, and international waters. In March 1994, after 18

287

Box 4 UN Convention on the Law of the Sea (UNCLOS)

In 1982, following a lengthy 24-year negotiation process, 159 nations signed the UN Convention on the Law of the Sea. The Convention puts in place a broad framework for protecting the marine environment. It essentially gives a binding effect to Principle 21 of the Stockholm Conference by requiring all states to ensure that pollution from activities under their jurisdiction or control does not cause damage to the marine environment of other states. In a broad sense, UNCLOS calls on states to prevent, reduce, and control pollution from land-based sources, the atmosphere, dumping, vessels and installations used in exploring and mining the sea bed. UNCLOS outlines the general obligation of all states to take measures against pollution of the marine environment, including the establishment of national rules, standards, recommended practices and procedures to achieve its objectives. States are also required to cooperate with neighbouring nations to harmonize policies and programmes at the appropriate regional level and to monitor, evaluate, and analyse effects of marine pollution.

While UNCLOS does not specifically address comprehensive protection of living resources of the marine environment, it does establish basic obligations of states to protect, conserve, and manage these resources in various zones within and beyond national jurisdiction. Within continental shelves and exclusive economic zones, individual coastal states have the power to determine allowable catch of living resources and UNCLOS reaffirms that states have sovereign rights to exploit these resources. It also defines the rights and obligations of states fishing on the high seas and calls for future international agreement on fishing.

Given 24-year negotiation process for UNCLOS and its 12-year period before coming into force, addressing environmental issues of marine waters is certainly controversial. While it is a "global convention" like the CBD and UNFCCC, UNCLOS is simply a framework convention without a funding mechanism for its pollution and fishery provisions and with no specific provisions or protocols for driving action. It is merely an umbrella for providing guidance on a general level for future action. In fact, it sets the stage for future negotiations between the North and South by preserving the ability of states to respond locally or regionally rather than in response to uniform global standards or requirements that might stunt national growth in the South. Despite this broad global legal framework and flexibility, the fact remains that the oceans and exclusive economic zones are being perilously overfished and coastal zones are purposely being used as receptacles for waste from land-based economic activities and communities.

Box 5 **Global Programme of Action for the Protection of the Marine Environment from Land-Based Activities (GAP)**

While land-based sources of marine pollution are specifically mentioned in two sections of UNCLOS (Articles 207 and 213), the lack of consensus and hesitancy over costs of commitments resulted in very general wording back in 1982. The general provisions about establishing guidelines, rules, and monitoring systems as well as harmonizing policies at the regional level (while taking into account the economic capacity of developing states and their need for economic development) did not result in progress and in fact coastal/marine ecosystems have become even further degraded over the last two decades.

In 1983, UNEP convened a group of experts to prepare guidelines (1985 Montreal Guidelines for the Protection of the Marine Environment Against Pollution from Land-Based Sources) to give life to these general provisions of UNCLOS. Unfortunately, the UNEP Governing Council only took note of the Guidelines rather than adopting them. As a result of planning for the Earth Summit and a 1991 meeting of experts held on the subject by UNEP, the subject was included in the negotiations of Chapter 17 of Agenda 21 and in fact was expanded from just pollution to all "land-based activities" such as physical habitat destruction, sediment, and airborne pollutants.

Chapter 17 of Agenda 21 called upon UNEP to convene an intergovernmental meeting on land-based activities, and Decision 17/20 of the UNEP Governing Council in May 1993 authorized the preparation process for adopting a programme of action. Following preparatory meetings in Montreal, Reykjavik, and Washington, 109 countries met at an intergovernmental meeting in Washington, D.C. in late 1995 to adopt the GPA and its "Washington Declaration."

In adopting the GPA, governments set as their common goal sustained and effective action to deal with all land-based imports on the marine environment. The GPA identifies actions needed at various levels of society to prevent, control, and reduce the degradation of the marine environment. It classifies areas of concern (such as sewage or persistent organic pollutants), establishes priority action areas and defines strategies and programmes to take advantage of numerous instruments that currently exist. While most financial resources must come from domestic sources, external sources such as bilateral donors and international financing institutions are also expected to cooperate. Unlike the UNFCCC and the CBD, the GPA does not include an explicit

---

Box 5 **(cont.)**

financial mechanism. However, the GEF is invited to build upon the work of the GPA and to fund the agreed incremental costs of activities consistent with GEF's Operational Strategy. An information clearinghouse mechanism is also included in the GPA to provide assistance to countries and UNEP is directed to support the regional thrust of the GPA through the Regional Seas Programme.

States declare their intention to: develop or review national programmes, implement the programmes, cooperate to build capacity and mobilize resources, take immediate preventive and remedial actions, cooperate on a regional basis, urge national and international institutions and bilateral donors to accord priority to projects under the GPA, and develop a global, legally binding instrument for the reduction or elimination of 12 persistent organic pollutants that are known to bioaccumulate in marine life and pose threats to human and ecosystem health. Various objectives, national activities, regional actions, and international actions are recommended for different areas of concern.

---

months of negotiations, agreement was reached in Geneva to transform the GEF from its pilot phase into a permanent financial mechanism. The restructured facility, with its $2bn trust fund, is open to universal participation (currently 155 countries) and builds upon the partnership between the United Nations Development Programme (UNDP), the United Nations Environment Programme (UNEP), and the World Bank – which are its implementing agencies. In addition to the four focal areas, activities to address land degradation are also eligible for funding insofar as they relate to one or more of the four focal areas.

In restructuring the GEF, governments ensured that it fully embodied the principles that were set out in the Rio conventions as well as Agenda 21. The GEF serves as a mechanism for international cooperation for the purpose of providing new and additional grant and concessional funding to meet the agreed incremental costs of measures that achieve global environmental benefits in the four focal areas. In October 1995, the GEF Council adopted an operational strategy, which represents the strategic framework for actions of the GEF in its four focal areas. According to the strategy's principles, the GEF will fund projects and programmes that are country-driven and based on national priorities designed to support sustainable development.

GEF operational strategy

The GEF operational strategy (GEF, 1996a) has been developed to guide the preparation of country-driven initiatives in the GEF's four focal areas. This strategy will assist the GEF Secretariat and its three implementing agencies in developing work programmes, business plans, and budgets. It will also guide the GEF Council in approving these activities.

The operational strategy for biodiversity sets forth an approach for implementing the GEF's mandate in biodiversity. It provides a framework for the development and implementation of GEF-financed activities to allow recipient countries to address the complex global challenge of biodiversity conservation and sustainable use. It also provides a framework for systematic monitoring and evaluation of the effectiveness of GEF-financed activities and incorporates guidance from the Conference of the Parties (COP) of the CBD.

The main strategic considerations guiding GEF-financed activities to secure global biodiversity benefits are: (a) integration of the conservation and sustainable use of biodiversity within national and, as appropriate, subregional and regional sustainable development plans and policies; (b) helping to protect and sustainably manage ecosystems through targeted and cost-effective interventions; (c) integration of efforts to achieve global benefits in other focal areas where feasible, and in the cross-sectoral area of land degradation, primarily desertification and deforestation; (d) development of a portfolio that encompasses representative ecosystems of global biodiversity significance; and (e) that GEF activities will be targeted and designed to help recipient countries achieve agreed biodiversity objectives in strategic and cost-effective ways.

The GEF operational strategy in climate change incorporates the policy guidance of the COP to the FCCC. The COP provided initial guidance on eligibility criteria, programme priorities, and policies for the financial mechanism, whose operation, on an interim basis, is entrusted to the GEF. Enabling activities facilitate implementation of effective response measures. Mitigation measures reduce or lead to the reduction of greenhouse gas emissions from anthropogenic sources or protect or enhance removal of such gases by sinks (thus reducing the risk of climate change). The GEF assists in implementation of national programmes by supporting agreed mitigation activities that meet either long-term or short-term criteria. Adaptation activities minimize the adverse effects of climate change. Initially, the GEF will meet the "agreed full costs of relevant adaptation activities undertaken in the context of the formulation of national communications." These are the "Stage I adaptation activities" outlined by the COP. Funding for adaptation activities beyond Stage I will be dependent on COP guidance. The overall strategic thrust of GEF-financed climate change activities is to support sustainable measures that minimize climate change damage by reducing the risk, or the adverse effects, of climate change. The GEF will finance agreed and eligible enabling, mitigation, and adaptation activities in eligible recipient countries.

Three initial operational programmes for the climate change focal area are proposed on the basis of a review of technical assessments, including recent work for the GEF on the cost reductions expected in new energy technologies. These programmes are consistent with the guidance provided by the COP and with the most recent findings of the IPCC. The three operational programmes that will be developed initially are:

– removal of barriers to energy conservation and energy efficiency;
– promotion of the adoption of renewable energy by removing barriers and reducing implementation costs;
– reduction of the long-term costs of low greenhouse gas-emitting energy technologies.

In the international waters area, GEF's objective is to contribute primarily as a catalyst to the implementation of a more comprehensive, ecosystem-based approach to managing international waters and their drainage basins as a means of achieving global environmental benefits. The GEF implementing agencies assist countries to find means of collaborating with neighbouring countries in order to change the ways human activities are undertaken in different economic sectors so that transboundary conflicts and problems can be resolved. The goal is to help groups of countries use the full range of technical, economic, financial, regulatory, and institutional measures needed to operationalize sustainable development strategies for transboundary water bodies and their contributing drainage basins.

The operational strategy (GEF, 1996a) outlines priorities to be addressed in this focal area. GEF activities focus on threatened transboundary water bodies and the most imminent threats to their ecosystems. Five types of action are targeting these hazards:

- Control of land-based sources of pollution that degrade the quality of international waters. Prevention of releases of persistent toxic substances and heavy metals, as well as nutrients and sediments, into basins of international waters with rare and endangered species or unique ecosystems is of particular importance. A particularly high priority is placed on interventions to address persistent organic pollutants (POPs).
- Prevention and control of land degradation where transboundary environmental concerns result from desertification or deforestation.
- Prevention of physical and ecological degradation of critical habitats (such as wetlands, shallow waters, and reefs) that sustain biodiversity and provide shelter and nursing areas for threatened and endangered species.
- Improved management and control measures that better guide the exploitation of living and non-living resources and address such problems as overfishing or excessive withdrawal and diversion of freshwater from transboundary basins.
- Control of ship-based sources of chemical washings and non-indigenous species which are transferred in ballast water and can disrupt ecosystems or adversely affect human health.

In its first five years, 65 developing countries and those in economic transition have received funding to participate in GEF international waters projects. Table 2 provides a listing of transboundary freshwater river basin, lake basin, and large marine ecosystems that have received GEF funding. A wide variety of situations in all five

Table 2 **GEF international waters projects**

Transboundary river basins*	Transboundary lake basins*	Large marine ecosystems*
Danube (14)	Lake Victoria (3)	Gulf of Guinea (5)
Dneiper (3)	Lake Tanganika (4)	East Asian seas (9)
Bermejo (2)	Lake Malawi (3)	Black sea (6)
Okavango (3)	Lake Titicaca (2)	Mediterranean (18)
Tumen (4)	Lake Ohrid (2)	Gulf of Aqaba (3)
Aral sea basin (5)		Red sea (6)
Plata maritime front (2)		W Indian ocean (8)

GEF (1996b). Table includes projects underway or in preparation with GEF funding.
*Figure in ( ) is number of countries.

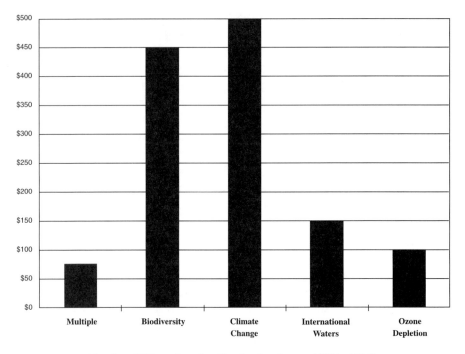

Fig. 2 **GEF project funding by focal area (1991–1996)**

economic development regions of the world are represented in the portfolio.

Over the first five years of GEF activities (three in a pilot phase and two following restructuring), approximately US$1.334bn in grant funding has been allocated to the focal areas as is shown in figure 2. Climate change projects have received approximately 39 per cent of the funding as of the end of 1996 while biodiversity projects received 35 per cent and international waters projects 12 per cent. Approximately 160 countries have expressed interest in officially participating with GEF.

A comprehensive approach to water resources

Over the last two decades, we have witnessed an evolution in water-resource management thinking – from the Mar del Plata Conference in the 1970s to the Law of the Sea Convention in the 1980s and the Dublin Statement and UNCED in the 1990s. This evolution has resulted in a consensus that a more comprehensive approach to water-resource management is needed – one that is cross sectoral in nature, that integrates ecological and developmental needs, and that is based on protecting the ecological sustainability of the water environment. This means that, at the highest levels of government, recognition should be given that water and watersheds must be managed as valuable natural resources to meet multiple uses rather than as mere inputs to specific sectoral activities.

The World Bank has recently called attention to mismanagement of surface and groundwater resources and the water environment as a significant impediment towards poverty reduction and sustainable development. The Bank's Water Resources Management Policy, adopted in 1993 after a lengthy process of consultation with NGOs, governments, and international organizations, calls attention to the need for countries and development organizations to adopt a more comprehensive approach to water-resources management (World Bank, 1993). This new approach represents a quantum shift from sector by sector projects to a more holistic approach recognizing the river basin as the appropriate unit for managing not only water quality, quantity, and ecosystems but also sectoral development initiatives. Economic sectors are now asked to take full responsibility for preventing the degradation of water resources by modifying existing activities, using pollution prevention strategies in new activities, and coordinating across sectors so that the water environment can be sustained for its multiple purpose uses. Interaction among different, but interrelated sectors is now being mandated so that sustainable development goals can be achieved. A whole host of financial management, economic, policy reform, technological, and participatory tools are also recommended including use of market-based instruments.

The Bank will provide assistance to member countries in developing a comprehensive approach to water-resources management suitable for the country's needs, resources, and capabilities. An emphasis is placed on building effective institutions to protect, enhance, and restore water quality and aquatic ecosystems that have been damaged by pollution or past development projects. Legal and regulatory reforms, emphasis on economic incentives, proper pricing policies, decentralization of water service deliveries, and active participation of beneficiaries, stakeholders, and the poor in water-resources management activities are stressed.

The comprehensive approach has important implications for involving different sectoral interests, different government ministries, and the public in participatory processes for improving water-resources management. Country water-resources management strategies (based on analysis prepared as part of cross-sectoral water-resources assessments) are being recommended as first steps for countries as they seek to implement this comprehensive approach. These processes provide opportunities to participate in identifying cross-sectoral influences on the water environment, finding opportunities for complementary cross-sectoral actions, setting priorities for action by different sectors, and then determining strategies for how to achieve sustainable, multiple use of the water environment. Such processes also provide the opportunity to build partnerships, institutional capacity, and NGO/stakeholder participation to more comprehensively manage a country's water resources in a sustainable fashion.

The GEF operational strategy also advocates this comprehensive approach for international waters projects (GEF, 1996a). Typically, projects begin with GEF-implementing agencies assisting the co-operating nations in undertaking strategic work. As noted in the operational strategy, this is done so that collaborating nations can each establish an inter-ministerial technical team to assemble information on the water-related environmental problems/conflicts and share this information with colleagues from the other nations in a committee setting. In this way, a transboundary water-resources analysis can be produced that contains the facts of the dispute, conflict, or problem as well as opportunities and needs. This factual analysis can serve as a start for determining country-driven environmental and water priorities. It also allows very complex basin problems to be divided into smaller, more manageable ones, each with a specific action programme for resolution.

As part of the process, the countries determine what actions, policy changes, regulatory developments, and sectoral programmes are needed to resolve the priority problems, threats, or conflicts. These steps allow for harmonization of actions among nations so that economic advantages do not accrue. The resulting GEF project provides an opportunity for the collaborating countries to diffuse political issues by focusing on technical fact-finding. They can learn about their shared transboundary ecosystem, learn how their sectoral activities and policies impact the water system, and learn to work together in joint problem solving – all without legalistic commitments and in a spirit of pragmatic cooperation because each country has a stake in the water issues. This joint effort may be able to build a sense of trust among participating individuals, and the experience may eventually lead to more formal and sustainable legal frameworks among nations in order to keep the initiative moving after the GEF project.

GEF's catalytic role helps to integrate transboundary water issues into national development plans, encourages the transfer of environmentally sound technology and knowledge and helps to strengthen the capacity of developing countries to play their full part in implementing needed interventions in different sectors. In essence, the GEF helps nations put together the essential pieces of a more comprehensive, ecosystems-based approach for managing transboundary waters as a means to operationalize sustainable development. GEF funds the transaction costs of these processes, it leverages the participation of other programmes and forms of development assistance, and it provides links to other GEF focal areas so that countries can effectively set priorities to achieve multiple benefits of GEF interventions.

Perspectives from the São Francisco basin

With its drainage area comparable to the Colorado or the Columbia rivers of North America, and with its economic importance to Brazil, the São Francisco basin is certainly one of the key basins of Latin America. Characterized by large cities like Belo Horizonte as well as by small villages in any of its six states, the basin holds about 16 per cent of the population of Brazil and flows for 2,700 kilometres before discharging to the Atlantic Ocean.

The World Bank has had a special role in the development of the São Francisco basin, which is known as the "River of National Unity." The basin has been the subject of development efforts for over 50 years. Government interventions, often constructed with loans from the World Bank, have been largely sectoral in nature and not integrated with environmental needs. The construction of hydropower projects and large irrigation schemes has been reported elsewhere in this volume as creating complex environmental problems. Complex changes in flows and timing in addition to competition for water among industrial, agricultural and hydropower sectors has created the need to adopt a more comprehensive and sustainable approach to development as part of addressing the environmental problems.

An excellent description of the basin is presented by Romano and Cadavid Garcia (this volume). Other key sources of information include an extensive analysis of lessons learned by the World Bank on environmental issues in the basin (World Bank, 1992) as well as an assessment of pollution loading from the basin to the Atlantic (Jennerjahn et al., 1996). Of great importance to Brazil is the significant role that the São Francisco valley has played in the development of the impoverished and drought-prone north-east and that of the Brazilian economy as a whole. Many World Bank assisted projects have helped Brazil reduce the widening income disparity between the north-east and south-central portions of Brazil and to combat rural poverty. The over 12,000 megawatts of installed hydropower capacity has powered industrial development in Brazil, and the water for irrigation in the semi-arid area has helped tackle rural poverty.

Environmental status

Beginning with the formation of the Companhia Hidro Eletrica do Sao Francisco (CHESF) in 1945 and the first World Bank loan to CHESF in 1950 to build the Paulo Afonso I power project, a number of governmental interventions in various sectors have been undertaken to promote regional development. The development was not planned in an integrated way, and it responded to market-driven exploitation as in other countries. Complex environmental problems have resulted worldwide from the traditional sector-based economic development strategies. As noted in the cited references, environmental deterioration, desertification, contamination of reservoirs, soil erosion, agricultural intensification, and drastically altered flow regimes have resulted in environmental degradation in the basin.

In the upper São Francisco basin, in the vicinity of the city of Belo Horizonte, contamination has resulted from the discharges of mines and ore-procession facilities, food-processing facilities, chemical and agrochemical industries, and discharges of sewage. Deforestation was also caused by logging for charcoal to fuel the industries. Consequences in the middle and lower middle São Francisco basin, due to dam construction, include destabilized river banks, which have contributed to increased movement of sediments downstream, changes in fisheries, eutrophication resulting from wastewater discharges, and land degradation (i.e. erosion and salination) due to agricultural development on marginal soils. In the lower São Francisco, environmental consequences include contamination from both agricultural and agro-industrial development, as well as wastewater discharges, which have contributed to public health concerns in the coastal zone.

Throughout the basin, the sectoral nature of the development along the São Francisco river has resulted in competition for water resources among the various industrial, agricultural, and hydroelectric production sectors in the basin. Soil degradation increases progressively from the headwaters to the delta, from 8.4 million tons/year at Pirapora to 32 million tons/year and more at Posto de Morpara and Manga. A significant proportion of this load enters the South-West Atlantic LME and is deposited in the São Francisco estuary, increasing marine algae production which contributes to eutrophication. Of even greater consequence was the creation of polders to dike coastal wetlands for agricultural use. With loss of wetlands due to the conversion and with changed flow regions, which eliminate cleansing floods that keep salt water closer to the ocean, complex environmental changes can occur like saltwater intrusion, accelerated sedimentation, pollution, loss of wetlands, and reduction of fisheries.

World Bank-assisted investments

The São Francisco is of critical importance to the survival of many small farmers in the impoverished north-east because it is the region's only perennial waterway. Without water there is no development, no biological diversity, no human existence – it is that simple. Development of the basin proceeded in a fragmented sector by sector, project by project manner just like the Columbia river, the Colorado river, the Ganges, and the Nile. As the World Bank (1992) acknowledged, this was a shortsighted approach that resulted in displacing almost 200,000 people from their lands and homes and has created complex changes in flow regimes and subsequent environmental quality.

Table 3 outlines the variety of Bank-assisted projects in the basin totalling billions of dollars. They run the range from dam construction, resettlement of displaced people, conversion of coastal wetlands to agricultural polders, hydropower projects, various traditional as well as quite modern irrigation projects, reforestation, and pollution abatement projects. Construction of Sobradinho dam and the Paulo Afonso cascade of power projects resulted in the increase of minimum flows at the coast that flooded out farmers who were encroaching on the wetlands. This flow increase prompted construction of dikes, polders, and irrigation schemes on the coastal converted wetlands to allow agricultural use. With increases in irrigation projects over time and with the desire to divert significant amounts of water

Table 3   **Selected world bank-assisted projects, São Francisco basin**

Project	Year approved
Paulo Afonso hydropower	1950
Paulo Afonso IV hydropower (Sobradinho dam)	1974
Lower São Francisco polders	1975
Lower São Francisco second irrigation project	1979
Upper and middle São Francisco irrigation project	1987
Irrigation subsector project	1988
Jaiba irrigation project	1988
Northeast irrigation I	1990
Minas Gerais forestry development	1988
Water supply and sewerage – Minas Gerais I	1974
Water supply and sewerage – Minas Gerais II	1976
Water supply and sewerage – Minas Gerais III	1980
Minas Gerais water quality and pollution control	1993

from the valley to the semi-arid north for development, further threats may result to the coast.

In the case of land degradation, loss of biodiversity at the coast as well as in drylands, and land-based activities that degrade marine waters, all the problems are globally significant issues. In particular, the beach area north of the delta represents critical habitat for threatened and endangered sea turtles. Indigenous peoples from 18 different groups also live in the region. Their existence is at question with much of the wildlife and fish they used to depend on for animal protein being in jeopardy. Pollution at the coast accumulated from the basin, deforestation, loss of flooding due to regulation may all threaten reproduction of fish and may affect tribal lifestyles.

Comprehensive basinwide approaches

Increasingly competitive demands for water use from different sectors range from hydropower to irrigation, navigation, pollution disposal, interbasin transfer, urban/industrial water supply, and upstream flow needs to sustain the water environment, its biodiversity, and indigenous people who depend on the water. In addition, existing environmental degradation from fragmented sectoral development also needs to be addressed.

Such challenges are common worldwide and even greater conflicts with more environmental damage have resulted in the Colorado, Columbia, Ganges, and Nile basins. Such unsustainable development

300

is inconsistent with the intent of Agenda 21 and the initiatives arising from the Earth Summit. However, there is no standard way to fix these problems or to incorporate global environmental interests into sectoral activities. Each convention seems to have its own separate requirements, initiatives and partners which invariably exclude sectoral ministries in governments and subnational levels of government.

The way ahead is through a more comprehensive, cross-sectoral, environmentally based approach to water resources management basin by basin as was advocated by Biswas (1993), World Bank (1993), and the GEF Operational Strategy (GEF, 1996a). This approach provides a way for different ministries and different subnational governments to participate in assessing water and land-related needs, problems, and opportunities in a basin and then work with stakeholders (even consult with them) to evaluate options, all opportunities, win-win situations from an environmental and an economic perspective to map a series of pragmatic interventions. This approach breaks the larger problems down into smaller, more manageable components for coordinated, practical action by different sectoral interests.

El-Ashry (1993), the World Bank (1993) in its water policy, and the GEF (1996a) in its Operational Strategy all encourage a balanced policy, regulatory, and demonstration approach to solving each component problem. Economic instruments are stressed and in fact are a key element in attempts to restore the Colorado (Morrison et al., 1996). Using these instruments not only for quantity purposes but also for water quality at the same time is an important concept. The US, Mexico, and Chile all use water-quantity markets, but this creates externalities and problems with the environment that need to be addressed simultaneously, not in a piecemeal way. Young and Congdon (1994) in particular describe how water-quality improvement markets can help address agriculture pollution.

Globally significant opportunities

As Mattos do Lemos (1996) has indicated, there is a new dawn of environmentalism in Brazil. Policies are changing and investments are being made in the environment. The São Francisco valley is a good example with a series of sewage and industrial pollution abatement prospects aimed at the Belo Horizonte area as noted in table 3 (Minas Gerais projects). An additional loan was requested to foster revegetation of cut-over and mined areas in the basin. This action provides global environment benefits by reducing erosion/land degradation and achieving sequestration of carbon. There has been a new approach to water resources in Brazil since January 1995 in bringing water-resources management under one umbrella. There is also a Citizens for Water Movement and a revitalized National Coastal Management Program resulting from an important meeting in Natal in 1996.

In the São Francisco, Romano and Cadavid Garcia (this volume) recognize that this comprehensive approach is needed to resolve conflicts, address important environment issues, and provide for sustainable development. The institutional element of this approach is important as well. A basin committee or basin organization that might be able to coordinate activities across sectors and levels of subnational government is needed and it would also facilitate private sector and other stakeholder participation. Economic instruments show great promise, but only if combined with environmental regulatory requirements to ensure a level playing field among industries. Other economic aspects of the comprehensive approach are evident in Bank-assisted irrigation projects such as the Formaso A and H Schemes with decentralized responsibilities of user groups and suitable fees and tariffs to recover costs.

Pilot activities to protect biodiversity, to test out how economic instruments for both water quantity and quality concerns could work, and actions to address land degradation would all be encouraging. Subnational government and other local interest is growing. An Inter-State Parliamentary Commission for the Development of the Rio São Francisco (CIPE) was formed that is comprised of the Presidents of the Legislative Assemblies of the five states comprising the largest portion of the land area of the basin. In addition, the local government authorities created UNIVALE, the Uniao das Prefeituras do Vale do São Francisco, which includes representation from the municipalities in the basin. This union provides technical advice on issues such as energy production, irrigation development, sanitation and human settlements, tourism, transportation, education, and environmental protection. Even industries in all six states have joined to participate in dialogue. The stage is being set to implement the comprehensive approach.

If wetlands at the coast are restored by pulling out some polders and dikes, if areas are reforested to address soil erosion, if drylands biodiversity and special turtle habitat are protected, and if basinwide hydropower flow regimes are altered to recreate limited floods that restore coastal wetlands for fish habitat for the indegeneous peoples, Brazil would have accomplished a globally significant activity. The flow regime is a key to ensure that water requirements to sustain coastal wetlands and fisheries is provided. Such actions would pilot an example of how the GPA might be implemented by UNEP to address land-based activities. While coastal and marine problems need to be diagnosed by marine specialists, remedial measures need to be cross sectoral in nature, need to be facilitated by freshwater basin experts, and must be undertaken in the context of new economic development fostered by comprehensive management.

Perspectives from the Plata basin

With a drainage area larger than the Nile, the massive Plata basin is second only to the Amazon in drainage area. Significant problems related to water pollution, deforestation, soil erosion and subsequent sedimentation, altered hydrologic regimes associated with increased downstream flooding, and coastal fisheries depletion have created a challenge for the five riparian nations. Cordeiro (this volume) summarizes the situation of the Plata basin quite well while Wetlands for the Americas (1993), Bonetto (1989), and Bisbal (1995) provide more detailed ecological information demonstrating significant degradation.

Global linkages

Given the transboundary nature of the basin, the existence of world class habitat for biodiversity in coastal wetlands, the Pantanal, and in the upper basin, needs for reforestation, and concerns about changing climate, Plata basin environmental issues are global in nature. The issues are globally significant and the pressures on the ecosystem have a global driving force. In addition, the pressures in one part of the Plata can have influence on land, people, and the global environment through deforestation in another region. In addition, certain policy choices can end up limiting the ability of a country, or its downstream neighbour, to adapt to changing global conditions. If these issues seem complex, it is because they are. These linkages have not only been identified in the Plata basin but also in the US in the 1970s in response to export markets, international monetary systems, and national policies designed with a single sector in mind (Cook, 1985).

The Paraná basin has numerous environmental problems resulting from over three dozen dams designed for single purpose hydropower (Bonetto, 1989). With this policy, very little storage is made available to control floods. As a result of increased oil prices, desire to gain foreign exchange from exports to service debt and increased world prices for soybeans, agricultural modernization was undertaken in several Brazilian states by the military government to grow soybeans intensively in the Paraná basin (Mahar, 1989). This involved large machinery which tilled the land and created massive soil erosion which ended up in the reservoirs. The small farmers were displaced by modernization and the World Bank assisted the government in opening the Amazon in the State of Rondonia to displaced people from the Paraná and the São Francisco. This resulted in the Amazon deforestation that alarmed the world in the 1970s and 1980s (Mahar, 1989). Even though hydropower is thought to be an alternative to reduce greenhouse gas emissions, Fearnside (1995) showed that reservoirs in Brazil can be significant contributors of greenhouse gases along with slash and burn deforestation. Complex linkages exist among transboundary water problems, biodiversity loss, emission of greenhouse gases, and international markets that dictate domestic policies.

As is described elsewhere, floods in the Plata from upstream areas have become more frequent and more severe. Four contributing factors may interact to create this situation: increased rainfall associated with the El Niño Southern Oscillation (ENSO), construction and operation of the series of 40 single purpose hydropower projects on the Paraná River, deforestation and conversion to soybeans in response to world market forces leading to serious soil erosion and increased stormflows, and downstream accumulations of sediment. Retrospective analyses are needed to determine primary causes of increased vulnerability to floods so that the feasibility of mitigation measures can be determined. If indeed climate change has warmed the ocean, influenced the ENSO, resulted in periodic droughts in the Brazilian north-east, and created more frequent floods in a basin with increased vulnerability due to its infrastructure, massive remedial measures would be needed in the basin. The massive downstream floods in 1983 and 1992 caused billions of dollars in damage to Argentina, Paraguay, and a small portion of Brazil. The World Bank provided an emergency loan for a $300m project for Argentina to fix damaged infrastructure after the 1992 flood and a subsequent $500m project in 1996 to raise dikes along the river, among other interventions. Globally significant changes may already be at work in the Plata basin.

Environmental threats

In addition to these complex influences, various transboundary water resources problems are evident. McCaffrey (1993) described disputes over dam construction in the Paraná basin between Brazil and Argentina. Excessive sedimentation from deforestation and over-grazing is significant in the Paraguay river basin from the Bermejo tributary system (Bolivia, Argentina). Toxic substances from mining and sedimentation is also a problem in the Pilcomayo (Bolivia, Paraguay, Argentina). In fact popular periodicals report releases of 400,000 tonnes of sludge loaded with toxic substances from mine tailings ponds. Complex water pollution problems from Buenos Aires are evident in the Plata estuary (toxic substances, bacteria, and eutrophication) and overfishing remains a serious concern at the coast (Bisbal, 1995).

Five years ago, Argentina, Brazil, Paraguay, and Uruguay signed a treaty aimed at creating a free-trade zone and customs union known as Mercosur. With the future addition of Bolivia and Chile and interest from the European Union, enormous monetary and trade pressures may create additional global stress in the basin. Again, the environmental linkages are complex with enormously expensive dredging needed to keep shipping lanes open as ship traffic increases. The toxic substances in dredged spoil may pose problems as would planned upriver navigation improvements known as the Hidrovia project. Significant threats to biodiversity in the world's largest wetland (Pantanal) and elsewhere on the navigation route have been raised by Wetlands for the Americas (1993).

Addressing global priority issues

While there is serious transboundary environmental stress as well as complex biodiversity loss and climate change issues to address in the Plata basin, action is beginning. Both Uruguay and Argentina had GEF biodiversity projects in the pilot phase. Bolivia and Argentina asked for and have received a GEF international waters project in the Bermejo to address land degradation, soil erosion, and sustainable development. Uruguay and Argentina now have funding approved to prepare a GEF international waters project for the lower Plata basin centring on the Plata maritime front and its pollution as well as fishery issues. Argentina also has a GEF biodiversity project to protect vulnerable forest lands in its protected areas system.

Mercosur itself may provide an opportunity to address these global issues. In the context of Mercosur, nations would need to harmonize their environmental protection regulations so unfair competition does not result. World class biodiversity habitat is still left and urban areas are targets for subsidy removal to enable pricing policies to assist in preventing greenhouse gas emissions and releases of sewage. Despite this, the basin is now in a state of increased vulnerability to climate change damage because of the deforestation, sedimentation, and lack of storage capacity in reservoirs. Clearly all five nations will need to work together in a comprehensive, cross-sectoral manner if this economically significant flooding problem is to be reduced.

The way ahead: Toward sustainable development

Today, the way we think about water goes to the very heart of the increasing worldwide concern about human health, the environment, and the path towards sustainable development. Of all the natural resources needed for economic development, water is the most crucial element. We have examined how global environmental issues relate to the São Francisco and Plata basins. Both for the single-country São Francisco basin and the five-country Plata basin, a more comprehensive, cross-sectoral approach to development is needed so that priorities may be set and existing problems may be fixed as part of new development initiatives. Reforestation and sustainable agricultural practices can help reduce greenhouse gas emissions. Existing wetlands can be protected to nurture biological diversity and areas that have been drained or diked can be restored to provide essential habitat. Transboundary water pollution problems can be corrected.

The enormous challenges outlined here are not limited to these five nations or even the developing world. The need to mainstream global environmental issues into economic development strategies is also critical for the North. Protecting the global environment is a complex undertaking that will require an unprecedented degree of cooperation between the North and the South. Protecting the global commons requires a global response.

It is a time of global change in other aspects as well. Development assistance is falling and private sector investment is growing. Globalization of markets, finance systems, information, and private sector speculation create strong forces with uncertain modalities for control. Even global conventions to address these common issues are in a state of flux and confusion: uncertainty and politics can sometimes stall action.

Because it is a time of change with no models for how to mainstream these important environmental issues, we have put forward the thesis that since each of these issues is linked to or influenced by water resources, a more comprehensive, cross-sectoral approach to water management might be a pragmatic way to test interventions, to experiment, and to determine priorities or critical sites for mainstreaming needed interventions. The river basin as an ecological unit can serve as a practical, fixed platform in a time of change for integrating the work of different sectors or for determining priorities so that interventions intended for one purpose might have multiple benefits for other purposes.

The strategy of adaptive management – of testing, learning, trying again – represents the only road forward. With declining funding and needed activities having a site-specific basis anyway, why not try to mainstream policies, activities, demonstration projects river basin by river basin with a goal of sustaining the ecological integrity of the water environment along with the goal of the particular intervention? Over time, governments will become more comfortable and experienced to the point where they may support needed sector-wide policy changes and activities.

Few issues have a greater impact on human life and the health of our planet than the way water resources are managed. If such a comprehensive approach proves useful, then an enormous opportunity exists for water professionals with skills in thinking cross sectorally and interdisciplinarily, in using their imagination, and in facilitating participation of stakeholders – especially the private sector – in undertaking priority-setting processes. A watershed provides a concrete way of thinking about abstract sectoral, policy, and globally significant interventions – a way of operationalizing conceptual notions. After enough testing and demonstrating during times of uncertainty on a solid platform of a river, lake, or coastal basin, mankind may just find that the road to sustainable development begins at home ... right on your own watershed!

References

Anderson, R.J., da Franca Ribeiro dos Santos, N., and Diaz, H.F. 1993. *An Analysis of Flooding in the Parana/Paraguay River Basin*, LATEN Dissemination Note no. 5. World Bank.

Bisbal, G.A. 1995. "The Southeast South American Shelf Large Marine Ecosystem." *Marine Policy* 19(1): 21–38.

Biswas, A.K. 1994. "Management of International Water Resources: Some Recent Developments." In: A.K. Biswas, ed. *International Waters of the Middle East*, pp. 185–214. Oxford: Oxford University Press.

—— 1993. "Water for Sustainable Development in the Twenty-First Century – A Global Perspective." In: Biswas, Jellali, Stout, eds. *Water for Sustainable Development in the Twenty-First Century*, pp. 7–17. Oxford: Oxford University Press.

Bonetto, A.A, 1989. "The Increasing Damming of the Panama Basin and Its Effects on the Lower Reaches." *Regulated Rivers* 4(4): 333–346.

Brown, L. 1996. *Tough Choices – Facing the Challenge of Food Security*. NY: Norton.

Clark, E.H., Hawerkamp, J.A., and Chapman, W. 1985. *Eroding Soils: The Off-Farm Impacts*. Conservation Foundation.

Cook, K.A. 1985. *Eroding Eden*. Roosevelt Center for American Policy Studies.

Cordeiro, N. 1998. "Environmental management issues in the Plata basin," this volume.

Duda, A.M. 1994. "Achieving pollution prevention goals for transboundary waters through international joint commission processes." *Water Science and Technology* 30(5): 223–231.

El-Ashry, M. 1993. "Policies for Water Resources Management in Arid and Semi-arid Regions." In: Biswas, Jellali, Stout, eds. *Water for Sustainable Development in the Twenty-First Century*, pp 7–17. Oxford: Oxford University Press.

Fearnside, P.M. 1995. "Hydroelectric Dams in the Brazilian Amazon as Sources of Greenhouse Gases." *Environmental Conservation* 22(1): 7–19.

Fell, N. 1996. "Outcasts from Eden." *New Scientist* 151(2045): 24–27.

Global Environment Facility. 1996a. *GEF Operational Strategy*. Washington, D.C.

—— 1996b. *Quarterly Operational Report*, November. Washington D.C.

Houghton, R.A. 1994. "The World-Wide Extent of Land-Use Change." *Bioscience* 44(5).

Jennerjahn, T.C., Ittekkot, V., and Carvalho, C.E.V. 1996. "Preliminary Data on Particle Flux off the São Francisco River, Eastern Brazil." In: V. Ittekkot, P. Schafer, S. Honjo, and P.J. Depetris, eds. *Particle Flux in the Ocean Scope 51*. London: Wiley.

Kindler, J., and Lintner, S. 1993. "An Action Plan to Clean Up the Baltic Sea." *Environment* 35(8): 7–15.

Mattos de Lemos, H. 1996. "Water Resources of the Amazon Basin." Luncheon Speech at AWRA 31st Annual Conference, 1995. Reprinted in *Hydata*, March 1996, 7–9.

Mahar, D.J. 1989. *Government Policies and Deforestation in Brazil's Amazon Region*. Washington, D.C.: World Bank.

McCaffrey, S. 1993. "Water, Politics, and International Law." In: P.H. Gleick, ed. *Water in Crisis*. New York: Stockholm Environment Institute, Oxford University Press.

Morrison, J.I., Postel, S.L., and Gleick, P.H. 1996. *The Sustainable Use of Water in the Lower Colorado River Basin.* Pacific Institute for Studies in Development, Environment, and Sustainability.

Postel, S. 1992. *Last Oasis – Facing Water Scarcity.* NY: Norton.

Romano, P.A., and Cadavid Garcia, E.A. 1998. "Policies for water-resources planning and management of the São Francisco river," this volume.

Roodman, D.M. 1996. *Paying the Piper: Subsidies, Politics, and the Environment,* World Watch Paper 133. World Watch Institute.

Tibbetts, J. 1996. "Farming and Fishing in the Wake of El-Nino." *Bioscience* 46(8): 566–569.

UN Environment Program (UNEP). 1996. *Global Biodiversity Assessment.* Cambridge: Cambridge University Press.

Wetlands for the Americas. 1993. *Hydrovia: An Initial Environmental Examination of the Paraguay-Panama Waterways.*

Wolfson, Z. 1990. "The Caspian Sea: Clear Signs of Disaster Environmental Policy." *Review for the Soviet Union and Eastern Europe* 4(2): 13–18.

World Bank. 1992. *Approach to the Environment in Brazil: A Review of Selected Projects,* vol. IV, Report no. 10039. Washington, D.C.

—— 1993. *Water Resource Management Policy Paper.* Washington, D.C.

Young, C.J., and Congdon, J.E. 1994. *Using Economic Incentives to Control Water Pollution from Agriculture.* Environmental Defense Fund.

Contributors

Asit K. Biswas
Chairman, Committee on International Collaboration,
International Water Resources Association,
Mexico, D.F. Mexico

Benedito P.F. Braga
President, International Water Resources Association,
São Paulo, Brazil

Eduardo Alfonso Cadavid Garcia
Researcher, Ministry of Water Resources,
Brasília, Brazil

Robin T. Clarke
Institute of Hydraulics Research,
Federal University of Rio Grande do Sul,
Porto Alegre, Brazil

Newton V. Cordeiro
Consultant, Organization of American States,
Washington, D.C., USA

Lilian del Castillo de Laborde
Director, Ministry of Foreign Affairs,
Buenos Aires, Argentina

Alfred M. Duda
Global Environment Facility,
Washington, D.C., USA

Fernando Genz
Institute of Hydraulics Research,
Federal University of Rio Grande do Sul,
Porto Alegre, Brazil

Haroldo Mattos de Lemos
Secretary, Ministry of Environment,
Brasília, Brazil

Manuel Picasso Botto
Deputy Secretary Pro-Tempore, Tratado de Cooperación Amazónica,
Lima, Peru

Víctor Pochat
Director General, Ministry of Water Resources,
Buenos Aires, Argentina

310

Paolo Alfonso Romano
Secretary, Ministry of Water Resources,
Brasília, Brazil

E. Salati
Foundation for Sustainable
Development,
Rio de Janeiro, Brazil

Larry D. Simpson
World Bank,
Washington, D.C., USA

Fabio Torrijos Quintero
Ambassador, Ministry of Foreign
Affairs,
Bogota, Colombia

Cecilia Tortajada
Royal Institute of Technology,
Stockholm, Sweden

Carlos E.M. Tucci
Professor, Institute of Hydraulics
Research, Federal University of Río
Grange do Sul,
Porto Alegre, Brazil

Reizo Utagawa
Managing Director, Nippon Foundation
Tokyo, Japan

Index

312